Continuous Delivery with Docker and Jenkins

Third Edition

Create secure applications by building complete
CI/CD pipelines

Rafał Leszko

BIRMINGHAM—MUMBAI

Continuous Delivery with Docker and Jenkins
Third Edition

Associate Group Product Manager: Rahul Nair

Publishing Product Manager: Niranjan Naikwadi

Senior Editor: Athikho Sapuni Rishana

Content Development Editor: Sayali Pingale

Technical Editor: Shruthi Shetty

Copy Editor: Safis Editing

Associate Project Manager: Neil Dmello

Proofreader: Safis Editing

Indexer: Tejal Daruwale Soni

Production Designer: Alishon Mendonca

Senior Marketing Coordinator: Sanjana Gupta

Marketing Coordinator: Nimisha Dua

First published: August 2017

Second edition: May 2019

Third edition: April 2022

Production reference: 1110422

Published by Packt Publishing Ltd.

Livery Place

35 Livery Street

Birmingham

B3 2PB, UK.

978-1-80323-748-0

www.packt.com

– To my wonderful wife, Maria, for all her love, wisdom, and smile.

Contributors

About the author

Rafał Leszko is a passionate software developer, trainer, and conference speaker living in Krakow, Poland. He has spent his career writing code, designing architecture, and being responsible for tech in a number of companies and organizations, including Hazelcast, Google, and CERN. Always open to new challenges, he has given talks and conducted workshops at numerous international conferences, including Devoxx and Voxxed Days.

About the reviewer

Werner Dijkerman is a freelance cloud (certified), Kubernetes, and DevOps engineer, currently focused on, and working with, cloud-native solutions and tools including AWS, Ansible, Kubernetes, and Terraform. He is also focused on Infrastructure as Code and monitoring the correct "thing" with tools such as Zabbix, Prometheus, and the ELK Stack, with a passion for automating everything and avoiding doing anything that resembles manual work.

– Big thanks, hugs, and shoutouts to Anca Borodi, Theo Punter, and everyone else at COERA!

Table of Contents

3

Configuring Jenkins

Section 2 – Architecting and Testing an Application

4

Continuous Integration Pipeline

5

Automated Acceptance Testing

6

Clustering with Kubernetes

Section 3 – Deploying an Application

7

Configuration Management with Ansible

8

Continuous Delivery Pipeline

9
Advanced Continuous Delivery

Best Practices

Assessments

Index

Other Books You May Enjoy

Preface

Continuous Delivery with Docker and Jenkins – Third Edition explains the advantages of combining Jenkins and Docker to improve the continuous integration and delivery process of app development. It starts with setting up a Docker server and configuring Jenkins on it. It then outlines the steps to build applications on Docker files and integrate them with Jenkins using continuous delivery processes such as continuous integration, automated acceptance testing, and configuration management.

Moving on, you will learn how to ensure quick application deployment with Docker containers, along with scaling Jenkins and using Kubernetes. After that, you will get to know how to deploy applications using Docker images and test them with Jenkins. Toward the end, the book will touch base with missing parts of the CD pipeline, which are the environments and infrastructure, application versioning, and non-functional testing.

By the end of the book, you will know how to enhance the DevOps workflow by integrating the functionalities of Docker and Jenkins.

Who this book is for

The book targets DevOps engineers, system administrators, Docker professionals, or any stakeholders who would like to explore the power of working with Docker and Jenkins together.

What this book covers

Chapter 1, Introducing Continuous Delivery, demonstrates the pitfalls of the traditional delivery process and describes success stories, including Amazon and Yahoo.

Chapter 2, Introducing Docker, provides a brief introduction to Docker and the concept of containerization and looks at the benefits in terms of running applications and services using this platform. In addition, we will also describe, step by step, how to set up Docker Community Edition on a local machine or a server running Linux and check to see whether Docker is running properly.

Chapter 3, Configuring Jenkins, introduces the Jenkins tool, their architecture, and procedures to install master/agent instances on a Docker server, without Docker and using Kubernetes. Then, we'll see how to scale agents. Finally, you will get a working Jenkins instance ready to build applications integrated with your source code repository service.

Chapter 4, Continuous Integration Pipeline, describes how the classic continuous integration pipeline entails three steps: checkout, building, and unit tests. In this chapter, you will learn how to build it using Jenkins and what other steps should be considered (such as code coverage and static code analysis).

Chapter 5, Automated Acceptance Testing, explains how, before releasing an application, you need to make sure that the whole system works as expected by running automated acceptance tests. Ordinarily, applications connect with databases, cache, messaging, and other tools that require other servers to run these services. This is why the whole environment has to be set up and kept ready before the test suite is started. In this chapter, you will learn Docker Registry concepts and how to build a system made of different components running as Docker containers.

Chapter 6, Clustering with Kubernetes, explains how to scale to multiple teams and projects using Docker tools. In this chapter, you will be introduced to Kubernetes and learn how to use it in the continuous delivery process.

Chapter 7, Configuration Management with Ansible, describes how, once you have scaled your servers, to deploy your application in production. In this chapter, you will learn how to release an application on a Docker production server using configuration management tools such as Chef and Ansible. Additionally, you will learn about the infrastructure as code approach and the Terraform tool.

Chapter 8, Continuous Delivery Pipeline, focuses on the missing parts of the final pipeline, which are the environments and infrastructure, application versioning, and non-functional testing. Once this chapter has been concluded, the complete continuous delivery pipeline will be ready.

Chapter 9, Advanced Continuous Delivery, explains how, after building a complete pipeline, you can address more difficult real-life scenarios. Beginning with parallelizing the pipeline tasks, we will then show how to roll back to the previous version, how to run performance tests, what to do with database changes, and how to proceed with legacy systems and manual tests.

Best Practices, this includes best practices to be followed throughout the book.

To get the most out of this book

Docker requires a 64-bit Linux operating system. All examples in this book have been developed using Ubuntu 20.04, but any other Linux system with kernel version 3.10 or above is sufficient.

Software/hardware covered in the book	OS requirements
Java 8+	Windows, macOS X, or Linux
Docker	
Jenkins	
Python	
Kubernetes	
Ansible	
Terraform	

Download the color images

We also provide a PDF file that has color images of the screenshots/diagrams used in this book. You can download it here: `https://static.packt-cdn.com/downloads/9781803237480_ColorImages.pdf`.

Code in Action

Code in Action videos for this book can be viewed at `https://bit.ly/3NSEPNA`.

Download the example code files

You can download the example code files for this book from GitHub at `https://github.com/PacktPublishing/Continuous-Delivery-With-Docker-and-Jenkins-3rd-Edition`. In case there's an update to the code, it will be updated on the existing GitHub repository.

We also have other code bundles from our rich catalog of books and videos available at `https://github.com/PacktPublishing/`. Check them out!

Conventions used

There are a number of text conventions used throughout this book.

`Code in text`: Indicates code words in text, database table names, folder names, filenames, file extensions, pathnames, dummy URLs, user input, and Twitter handles. Here is an example: "We can create a new pipeline called `calculator` and, as pipeline script, put the code in a stage called `Checkout`."

A block of code is set as follows:

```
pipeline {
    agent any
    stages {
        stage("Checkout") {
            steps {
                git url: 'https://github.com/leszko/
calculator.git', branch: 'main'
            }
        }
    }
}
```

Any command-line input or output is written as follows:

```
$ sudo apt-get update
```

Bold: Indicates a new term, an important word, or words that you see onscreen. For example, words in menus or dialog boxes appear in the text like this. Here is an example: "Select **Gradle Project** instead of **Maven Project** (you can choose Maven if you prefer it to Gradle)."

> Tips or Important Notes
> Appear like this.

Get in touch

Feedback from our readers is always welcome.

General feedback: If you have questions about any aspect of this book, mention the book title in the subject of your message and email us at `customercare@packtpub.com`.

Errata: Although we have taken every care to ensure the accuracy of our content, mistakes do happen. If you have found a mistake in this book, we would be grateful if you would report this to us. Please visit www.packtpub.com/support/errata, selecting your book, clicking on the Errata Submission Form link, and entering the details.

Piracy: If you come across any illegal copies of our works in any form on the Internet, we would be grateful if you would provide us with the location address or website name. Please contact us at copyright@packt.com with a link to the material.

If you are interested in becoming an author: If there is a topic that you have expertise in and you are interested in either writing or contributing to a book, please visit authors.packtpub.com.

Share Your Thoughts

Once you've read *Continuous Delivery with Docker and Jenkins*, we'd love to hear your thoughts! Scan the QR code below to go straight to the Amazon review page for this book and share your feedback.

https://packt.link/r/1803237481

Your review is important to us and the tech community and will help us make sure we're delivering excellent quality content.

Section 1 – Setting Up the Environment

In this section, we will be introduced to Docker, and we will cover concepts such as continuous delivery and its benefits, as well as containerization. Furthermore, we will also be introduced to the Jenkins tool, and the architecture and procedures required to install master/slave instances on a Docker server, without Docker, and using cloud environments.

The following chapters are covered in this section:

- *Chapter 1, Introducing Continuous Delivery*
- *Chapter 2, Introducing Docker*
- *Chapter 3, Configuring Jenkins*

1

Introducing Continuous Delivery

A common problem that's faced by most developers is how to release the implemented code quickly and safely. The delivery process that's traditionally used is a source of pitfalls and usually leads to the disappointment of both developers and clients. This chapter will present the idea of the **continuous delivery** (**CD**) approach and provide the context for the rest of this book.

In this chapter, we will cover the following topics:

- Understanding CD
- The automated deployment pipeline
- Prerequisites to CD
- Combining CD and microservices
- Building the CD process

Understanding CD

The most accurate definition of CD is stated by *Jez Humble* and reads as follows:

> *"Continuous delivery is the ability to get changes of all types – including new features, configuration changes, bug fixes, and experiments – into production, or into the hands of users, safely and quickly, in a sustainable way."*

This definition covers the key points.

To understand this better, let's imagine a scenario. You are responsible for a product – let's say, an email client application. Users come to you with a new requirement: they want to sort emails by size. You decide that the development will take around 1 week. *When can the user expect to use the feature?* Usually, after the development is done, you hand over the completed feature to the **Quality Assurance** (**QA**) team and then to the operations team, which takes additional time, ranging from days to months.

Therefore, even though the development only took 1 week, the user receives it in a couple of months! The CD approach addresses this issue by automating manual tasks so that the user can receive a new feature as soon as it's implemented.

To help you understand what to automate and how, we'll start by describing the delivery process that is currently used for most software systems.

The traditional delivery process

The traditional delivery process, as its name suggests, has been in place for many years and is implemented in most IT companies. Let's define how it works and comment on its shortcomings.

Introducing the traditional delivery process

Every delivery process begins with the requirements that have been defined by a customer and ends with the product being released to production. There are differences between these two stages. Traditionally, this process looks as follows:

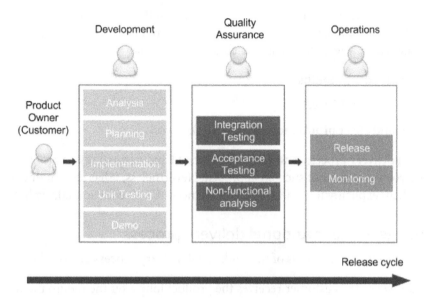

Figure 1.1 – Release cycle diagram

The release cycle starts with the requirements provided by the **Product Owner**, who represents the **Customer** (stakeholders). Then, there are three phases, during which the work is passed between different teams:

- **Development**: The developers (sometimes together with business analysts) work on the product. They often use agile techniques (Scrum or Kanban) to increase the development velocity and improve communication with the client. Demo sessions are organized to obtain a customer's quick feedback. All good development techniques (such as **test-driven development** (**TDD**) or extreme programming practices) are welcome. Once the implementation is complete, the code is passed to the QA team.

- **Quality Assurance**: This phase is usually called **User Acceptance Testing** (**UAT**) and it requires the code to be frozen on the trunk code base so that no new development will break the tests. The QA team performs a suite of **integration testing**, **acceptance testing**, and **non-functional analysis** (performance, recovery, security, and so on). Any bug that is detected goes back to the development team, so the developers usually have their hands full. After the UAT phase is completed, the QA team approves the features that have been planned for the next release.

- **Operations**: The final phase, and usually the shortest one, involves passing the code to the operations team so that they can perform the release and monitor the production environment. If anything goes wrong, they contact the developers so that they can help with the production system.

The length of the release cycle depends on the system and the organization, but it usually ranges from 1 week to a few months. The longest I've heard about was 1 year. The longest I worked on one was quarterly-based, and each part was as follows:

- **Development**: 1.5 months
- **UAT**: 1 month and 3 weeks
- **Release (and strict production monitoring)**: 1 week

The traditional delivery process is widely used in the IT industry, so this is probably not the first time you've read about such an approach. Nevertheless, it has several drawbacks. Let's look at them explicitly to understand why we need to strive for something better.

Shortcomings of the traditional delivery process

The most significant shortcomings of the traditional delivery process are as follows:

- **Slow delivery**: The customer receives the product long after the requirements were specified. This results in unsatisfactory time to market and delays customer feedback.

- **Long feedback cycle**: The feedback cycle is not only related to customers but developers. Imagine that you accidentally created a bug, and you learn about it during the UAT phase. *How long does it take to fix something you worked on 2 months ago?* Even dealing with minor bugs can take weeks.

- **Lack of automation**: Rare releases do not encourage automation, which leads to unpredictable releases.

- **Risky hotfixes**: Hotfixes cannot usually wait for the full UAT phase, so they tend to be tested differently (the UAT phase is shortened) or not tested at all.

- **Stress**: Unpredictable releases are stressful for the operations team. What's more, the release cycle is usually tightly scheduled, which imposes additional stress on developers and testers.

- **Poor communication**: Work that's passed from one team to another represents the waterfall approach, in which people start to care only about their part, rather than the complete product. If anything goes wrong, that usually leads to the blame game instead of cooperation.

- **Shared responsibility**: No team takes responsibility for the product from A to Z:

  - **For developers**: *Done* means that the requirements have been implemented.

  - **For testers**: *Done* means that the code has been tested.

  - **For operations**: *Done* means that the code has been released.

- **Lower job satisfaction**: Each phase is interesting for a different team, but other teams need to support the process. For example, the development phase is interesting for developers but, during the other two phases, they still need to fix bugs and support the release, which is usually not interesting for them at all.

These drawbacks represent just the tip of the iceberg of the challenges related to the traditional delivery process. You may already feel that there must be a better way to develop software and this better way is, obviously, the CD approach.

The benefits of CD

How long would it take your organization to deploy a change that involves just a single line of code? Do you do this on a repeatable, reliable basis? These are the famous questions from *Mary* and *Tom Poppendieck* (authors of *Implementing Lean Software Development*), which have been quoted many times by *Jez Humble* and others. The answers to these questions are the only valid measurement of the health of your delivery process.

To be able to deliver continuously, and not spend a fortune on the army of operations teams working 24/7, we need automation. That is why, in short, CD is all about changing each phase of the traditional delivery process into a sequence of scripts called the *automated deployment pipeline*, or the *CD pipeline*. Then, if no manual steps are required, we can run the process after every code change and deliver the product continuously to users.

CD lets us get rid of the tedious release cycle and brings the following benefits:

- **Fast delivery**: Time to market is significantly reduced as customers can use the product as soon as development is completed. Remember that the software delivers no revenue until it is in the hands of its users.

- **Fast feedback cycle**: Imagine that you created a bug in the code, which goes into production the same day. *How much time does it take to fix something you worked on the same day?* Probably not much. This, together with the quick rollback strategy, is the best way to keep production stable.

- **Low-risk releases**: If you release daily, the process becomes repeatable and much safer. As the saying goes, *if it hurts, do it more often.*

- **Flexible release options**: If you need to release immediately, everything is already prepared, so there is no additional time/cost associated with the release decision.

Needless to say, we could achieve all these benefits simply by eliminating all the delivery phases and proceeding with development directly from production. However, this would result in a reduction in quality. The whole difficulty of introducing CD is the concern that the quality would decrease alongside eliminating any manual steps. In this book, we will show you how to approach CD safely and explain why, contrary to common beliefs, products that are delivered continuously contain fewer bugs and are better adjusted to the customer's needs.

Success stories

My favorite story on CD was told by Rolf Russell at one of his talks. It goes as follows. In 2005, Yahoo! acquired Flickr, and it was a clash of two cultures in the developer's world. Flickr, by that time, was a company with the start-up approach in mind. Yahoo!, on the other hand, was a huge corporation with strict rules and a safety-first attitude. Their release processes differed a lot. While Yahoo used the traditional delivery process, Flickr released many times a day. Every change that was implemented by developers went into production the same day. They even had a footer at the bottom of their page showing the time of the last release and the avatars of the developers who made the changes.

Yahoo! deployed rarely, and each release brought a lot of changes that were well-tested and prepared. Flickr worked in very small chunks; each feature was divided into small incremental parts, and each part was deployed to production quickly. The difference is presented in the following diagram:

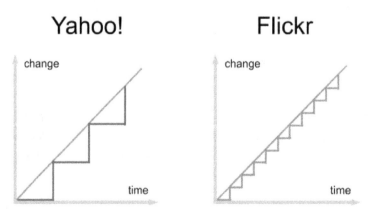

Figure 1.2 – Comparison of the release cycles of Yahoo! and Flickr

You can imagine what happened when the developers from the two companies met. Yahoo! treated Flickr's colleagues as irresponsible junior developers, a bunch of software cowboys who didn't know what they were doing. So, the first thing they wanted to do was add a QA team and the UAT phase to Flickr's delivery process. Before they applied the change, however, Flickr's developers had only one wish. They asked to evaluate the most reliable products throughout Yahoo! as a whole. It came as a surprise when they saw that even with all the software in Yahoo!, Flickr had the lowest downtime. The Yahoo! team didn't understand it at first, but they let Flickr stay with their current process anyway. After all, they were engineers, so the evaluation result was conclusive. Only after some time had passed did the Yahoo! developers realize that the CD process could be beneficial for all the products in Yahoo! and they started to gradually introduce it everywhere.

The most important question of the story remains: *how come Flickr was the most reliable system?* The reason behind this was what we already mentioned in the previous sections. A release is less risky if the following is true:

- The delta of code changes is small
- The process is repeatable

That is why, even though the release itself is a difficult activity, it is much safer when it's done frequently.

The story of Yahoo! and Flickr is only one example of many successful companies where the CD process proved to be the correct choice. Nowadays, it's common for even small organizations to release frequently and market leaders such as Amazon, Facebook, Google, and Netflix perform thousands of releases per day.

> **Information**
>
> You can read more about the research on the CD process and individual case studies at `https://continuousdelivery.com/evidence-case-studies/`.

Keep in mind that the statistics get better every day. However, even without any numbers, just imagine a world in which every line of code you implement goes safely into production. Clients can react quickly and adjust their requirements, developers are happy as they don't have to solve that many bugs, and managers are satisfied because they always know the current state of work. After all, remember that the only true measure of progress is the software that is released.

The automated deployment pipeline

We already know what the CD process is and why we use it. In this section, we'll describe how to implement it.

Let's start by emphasizing that each phase in the traditional delivery process is important. Otherwise, it would never have been created in the first place. No one wants to deliver software without testing it! The role of the UAT phase is to detect bugs and ensure that what the developers have created is what the customer wanted. The same applies to the operations team – the software must be configured, deployed to production, and monitored. That's out of the question. So, *how do we automate the process so that we preserve all the phases?* That is the role of the automated deployment pipeline, which consists of three stages, as shown in the following diagram:

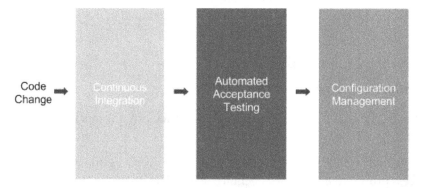

Figure 1.3 – Automated deployment pipeline

The automated deployment pipeline is a sequence of scripts that is executed after every code change is committed to the repository. If the process is successful, it ends up being deployed to the production environment.

Each step corresponds to a phase in the traditional delivery process, as follows:

- **Continuous integration**: This checks to make sure that the code that's been written by different developers is integrated.

- **Automated acceptance testing**: This checks if the client's requirements have been met by the developers implementing the features. This testing also replaces the manual QA phase.

- **Configuration management**: This replaces the manual operations phase; it configures the environment and deploys the software.

Let's take a deeper look at each phase to understand its responsibility and what steps it includes.

Continuous integration

The **continuous integration** (**CI**) phase provides the first set of feedback to developers. It checks out the code from the repository, compiles it, runs unit tests, and verifies the code's quality. If any step fails, the pipeline's execution is stopped and the first thing the developers should do is fix the CI build. The essential aspect of this phase is time; it must be executed promptly. For example, if this phase took 1 hour to complete, the developers would commit the code faster, which would result in a constantly failing pipeline.

The CI pipeline is usually the starting point. Setting it up is simple because everything is done within the development team, and no agreement with the QA and operations teams is necessary.

Automated acceptance testing

The automated acceptance testing phase is a suite of tests written together with the client (and QAs) that is supposed to replace the manual UAT stage. It acts as a quality gate to decide whether a product is ready to be released. If any of the acceptance tests fail, pipeline execution is stopped and no further steps are run. It prevents movement to the configuration management phase and, hence, the release.

The whole idea of automating the acceptance phase is to build quality into the product instead of verifying it later. In other words, when a developer completes the implementation, the software is delivered together with the acceptance tests, which verify that the software is what the client wanted. That is a large shift in thinking concerning testing software. There is no longer a single person (or team) who approves the release, but everything depends on passing the acceptance test suite. That is why creating this phase is usually the most difficult part of the CD process. It requires close cooperation with the client and creating tests at the beginning (not at the end) of the process.

> **Note**
>
> Introducing automated acceptance tests is especially challenging in the case of legacy systems. We will discuss this topic in more detail in *Chapter 9, Advanced Continuous Delivery*.

There is usually a lot of confusion about the types of tests and their place in the CD process. It's also often unclear as to how to automate each type, what the coverage should be, and what the role of the QA team should be in the development process. Let's clarify this using the Agile testing matrix and the testing pyramid.

The Agile testing matrix

Brian Marick, in a series of his blog posts, classified software tests in the form of the agile testing matrix. It places tests in two dimensions – business - or technology-facing – and supports programmers or a critique of the product. Let's have a look at this classification:

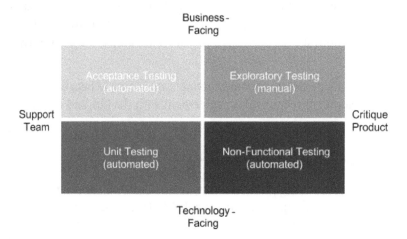

Figure 1.4 – Brian Marick's testing matrix

Let's look at each type of test:

- **Acceptance Testing (automated)**: These are tests that represent the functional requirements that are seen from the business perspective. They are written in the form of stories or examples by clients and developers so that they can agree on how the software should work.

- **Unit Testing (automated)**: These are tests that help developers provide high-quality software and minimize the number of bugs.

- **Exploratory Testing (manual)**: This is the manual black-box testing phase, which tries to break or improve the system.

- **Non-Functional Testing (automated)**: These are tests that represent system properties related to performance, scalability, security, and so on.

This classification answers one of the most important questions about the CD process: *what is the role of a QA in the process?*

Manual QAs perform exploratory testing, which means they play with the system, try to break it, ask questions, and think about improvements. Automation QAs help with non-functional and acceptance testing; for example, they write code to support load testing. In general, QAs don't have a special place in the delivery process, but rather have a role in the development team.

> **Note**
>
> In the automated CD process, there is no longer a place for manual QAs who perform repetitive tasks.

You may look at the classification and wonder why you see no integration tests there. *Where are they according to Brian Marick, and where can we put them in the CD pipeline?*

To explain this well, we need to mention that the meaning of an integration test differs based on the context. For (micro) service architectures, they usually mean the same as acceptance testing, as services are small and need nothing more than unit and acceptance tests. If you build a modular application, then integration tests usually mean component tests that bind multiple modules (but not the whole application) and test them together. In that case, integration tests place themselves somewhere between acceptance and unit tests. They are written in a similar way to acceptance tests, but they are usually more technical and require mocking not only external services but also internal modules. Integration tests, similar to unit tests, represent the code's point of view, while acceptance tests represent the user's point of view. In regards to the CD pipeline, integration tests are simply implemented as a separate phase in the process.

The testing pyramid

The previous section explained what each test type represents in the process, but mentioned nothing about how many tests we should develop. So, *what should the code coverage be in the case of unit testing? What about acceptance testing?*

To answer these questions, *Mike Cohn*, in his book *Succeeding with Agile*, created a so-called **testing pyramid**. The following diagram should help you develop a better understanding of this:

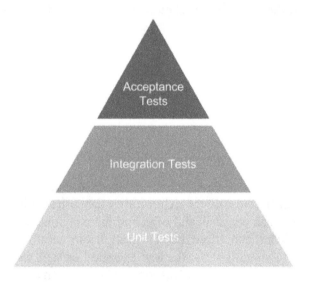

Figure 1.5 – Mike Cohn's testing pyramid

When we move up the pyramid, the tests become slower and more expensive to create. They often require user interfaces to be touched and a separate test automation team to be hired. That is why acceptance tests should not target 100% coverage. On the contrary, they should be feature-oriented and only verify selected test scenarios. Otherwise, we would spend a fortune on test development and maintenance, and our CD pipeline build would take ages to execute.

The case is different at the bottom of the pyramid. Unit tests are cheap and fast, so we should strive for 100% code coverage. They are written by developers, and providing them should be a standard procedure for any mature team.

I hope that the agile testing matrix and the testing pyramid clarified the role and the importance of acceptance testing.

Now, let's look at the last phase of the CD process: configuration management.

Configuration management

The configuration management phase is responsible for tracking and controlling changes in the software and its environment. It involves taking care of preparing and installing the necessary tools, scaling the number of service instances and their distribution, infrastructure inventory, and all the tasks related to application deployment.

Configuration management is a solution to the problems that are posed by manually deploying and configuring applications in production. This common practice results in an issue whereby we no longer know where each service is running and with what properties. Configuration management tools (such as Ansible, Chef, and Puppet) enable us to store configuration files in the version control system and track every change that was made to the production servers.

Additional effort to replace the operations team's manual tasks involves taking care of application monitoring. This is usually done by streaming the logs and metrics of the running systems to a common dashboard, which is monitored by developers (or the DevOps team, as explained in the next section).

One other term related to configuration management that has recently gained a lot of traction is **Infrastructure as Code** (**IaC**). If you use the cloud instead of bare-metal servers, then tools such as Terraform or AWS CloudFormation let you store the description of your infrastructure, not only your software, in the version control system. We will discuss both configuration management and IaC in *Chapter 7, Configuration Management with Ansible*.

Prerequisites to CD

The rest of this book is dedicated to technical details on how to implement a successful CD pipeline. The success of this process, however, depends not only on the tools we present throughout this book. In this section, we will take a holistic look at the whole process and define the CD requirements in three areas:

- Your organization's structure and its impact on the development process
- Your products and their technical details
- Your development team and the practices you adopt

Let's start with the organizational prerequisites.

Organizational prerequisites

The way your organization works has a high impact on the success of introducing the CD process. It's a bit similar to introducing Scrum. Many organizations would like to use the Agile process, but they don't change their culture. You can't use Scrum in your development team unless the organization's structure has been adjusted for that. For example, you need a product owner, stakeholders, and a management team that understands that no requirement changes are possible during the sprint. Otherwise, even with good intentions, you won't make it. The same applies to the CD process; it requires you to adjust how the organization is structured. Let's have a look at three aspects: the DevOps culture, a client in the process, and business decisions.

DevOps culture

A long time ago, when software was written by individuals or micro teams, there was no clear separation between development, quality assurance, and operations. A person developed the code, tested it, and then put it into production. If anything went wrong, the same person investigated the issue, fixed it, and redeployed it to production. The way the development process is organized changed gradually; systems became larger and development teams grew. Then, engineers started to become specialized in one area. This made perfect sense as specialization caused a boost in productivity. However, the side effect was the communication overhead. This is especially visible if developers, QAs, and operations are in separate departments in the organization, sit in different buildings, or are outsourced to different countries. This organizational structure is not good for the CD process. We need something better; we need to adopt the DevOps culture.

DevOps culture means, in a sense, going back to the roots. A single person or a team is responsible for all three areas, which are shown in the following diagram:

Figure 1.6 – DevOps culture

The reason it's possible to move to the DevOps model without losing productivity is automation. Most of the tasks that are related to QA and operations are moved to the automated delivery pipeline, so they can be managed by the development team.

> **Information**
>
> A DevOps team doesn't necessarily need to consist of only developers. A very common scenario in many organizations that are transforming is to create teams with four developers, one QA, and one person from operations. However, they need to work closely together (sit in one area, have stand-ups together, and work on the same product).

The culture of small DevOps teams affects the software architecture. Functional requirements have to be separated into (micro) services or modules so that each team can take care of an independent part.

> **Information**
>
> The impact of the organization's structure on the software architecture was observed in 1967 and formulated as Conway's law: "*Any organization that designs a system (defined broadly) will produce a design whose structure is a copy of the organization's communication structure.*"

A client in the process

The role of a client (or a product owner) changes slightly during CD adoption. Traditionally, clients are involved in defining requirements, answering questions from developers, attending demos, and taking part in the UAT phase to determine whether what was built is what they had in mind.

In CD, there is no UAT, and a client is essential in the process of writing acceptance tests. For some clients, who have already written their requirements in a testable manner, this is not a big shift. For others, it means changing their way of thinking to make the requirements more technically oriented.

> **Information**
>
> In the agile environment, some teams don't even accept user stories (requirements) without acceptance tests attached. These techniques, even though they may sound too strict, often lead to better development productivity.

Business decisions

In most companies, the business has an impact on the release schedule. After all, the decision of what features are delivered, and when, is related to different departments within the company (for example, marketing) and can be strategic for the enterprise. That is why the release schedule has to be re-approached and discussed between the business and the development teams.

There are techniques, such as feature toggles or manual pipeline steps, that help with releasing features at the specified time. We will describe them later in this book. To be precise, the term *continuous delivery* is not the same as *continuous deployment*. The latter means that each commit to the repository is automatically released to production. Continuous delivery is less strict and means that each commit ends up with a release candidate, so it allows the last step (from release to production) to be manual.

> **Note**
> Throughout the remainder of this book, we will use the terms continuous delivery and continuous deployment interchangeably.

Technical and development prerequisites

From the technical side, there are a few requirements to keep in mind. We will discuss them throughout this book, so let's only mention them here without going into detail:

- **Automated build, test, package, and deploy operations**: All operations need to be able to be automated. If we deal with a system that is non-automatable, for example, due to security reasons or its complexity, it is impossible to create a fully automated delivery pipeline.

- **Quick pipeline execution**: The pipeline must be executed promptly, preferably in 5-15 minutes. If our pipeline execution takes hours or days, it will not be possible to run it after every commit to the repository.

- **Quick failure recovery**: The possibility of a quick rollback or system recovery is necessary. Otherwise, we risk production health due to frequent releases.

- **Zero-downtime deployment**: The deployment cannot have any downtime since we release it many times a day.

- **Trunk-based development**: Developers must check into one main branch regularly. Otherwise, if everyone develops in their branches, integration is rare, which means that releases are rare, which is exactly the opposite of what we want to achieve.

We will learn more about these prerequisites and how to address them throughout this book. With this in mind, let's move to the last section of this chapter and introduce what system we plan to build in this book and what tools we will use for that purpose.

Combining CD and microservices

We live in the world of microservices. Nowadays, every system is either microservice-based or in the process of becoming microservice-based. After the first publication of the bestseller book by Sam Newman, *Building Microservices*, the software world has shifted into the fine-grained modular systems in which all communication happens over the network. Some companies have gone one step further and realized that they need to consolidate some of the microservices as they created too many of them. Some other companies even take a step back and consolidate microservices into a monolith.

While the topic of microservices is broad on its own and outside the scope of this book, it is important to understand how the microservice architecture affects the CD pipeline. Should we create a separate pipeline for each service? If yes, then how do we test the interaction between the services and the system as a whole?

Before answering these questions, let's look at the following diagram, which represents a small microservice-based system:

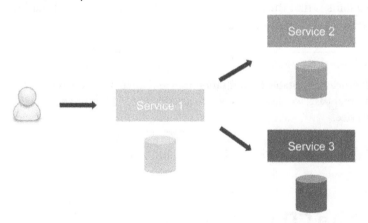

Figure 1.7 – Sample microservice system

There are three services in our system, each with a database. A user only interacts with *Service 1*. As a more concrete example, this system could represent an online store, where *Service 1* could represent the checkout service, *Service 2* could represent the product catalog service, and *Service 3* could represent customer service.

We could either implement one CD pipeline for the entire system or a separate CD pipeline for each microservice. Which approach is the right one? Let's consider both options. If we create one CD pipeline, this means that the automated acceptance testing phase runs against the entire system from the end user's perspective, which seems correct. However, one CD pipeline also means that we deploy all the services at the same time, which is completely against the microservice principles. Remember that in every microservice-based system, services are loosely coupled and should always be independently deployable.

So, we need to take the second approach and create a separate CD pipeline for each service. However, in such a case, the automated acceptance testing phase never runs against the entire system. So, how can we be sure that everything works correctly from the end user's perspective? To answer this question, we need a little more context about the microservice architecture.

In the microservice architecture, each service is a separate unit that's usually developed and maintained by a separate team. Services are loosely coupled, and they communicate over a well-defined API, which should always be kept backward compatible. In that context, each internal microservice does not differ much from an external service. That's why we should always be able to deploy a new service without testing other services. Note that it does not exclude the possibility of having a separate acceptance test for the entire system. All it explains is that the acceptance test of the entire system should not be a gatekeeper for deploying a single service.

> **Information**
>
> The CD process is suitable for both monolith and microservice-based systems. In the former case, we should always create a separate CD pipeline for each microservice.

For the sake of simplicity, all the examples in this book present a system that consists of a single service.

Building the CD process

So far, we've introduced the idea, benefits, and prerequisites concerning the CD process. In this section, we will describe the tools that will be used throughout this book and their place in the system as a whole.

> **Information**
>
> If you're interested in the idea of the CD process, have a look at the excellent book by *Jez Humble* and *David Farley*, called *Continuous Delivery: Reliable Software Releases through Build, Test, and Deployment Automation.*

Introducing tools

First of all, the tool is always less important than understanding its role in the process. In other words, any tool can be replaced with another one that plays the same role. For example, Jenkins can be replaced with Atlassian Bamboo, and Chef can be used instead of Ansible. This is why each chapter will begin with a general description of why such a tool is necessary and its role in the whole process. Then, the tool will be described in comparison to its substitutes. This will give you the flexibility to choose the right one for your environment.

Another approach could be to describe the CD process at the idea level; however, I strongly believe that giving an exact example, along with the code extract – something that you can run by yourself – results in a much better understanding of the concept.

> **Information**
>
> There are two ways to read this book. The first is to read and understand the concepts of the CD process. The second is to create an environment and execute all the scripts while reading to understand the details.

Let's take a quick look at the tools we will use throughout this book. This section, however, is only a brief introduction to each technology – more details will be provided later in this book.

The Docker ecosystem

Docker, as the clear leader of the containerization movement, has dominated the software industry in recent years. It allows us to package an application in an environment-agnostic image and treats servers as a farm of resources, rather than machines that must be configured for each application. Docker was a clear choice for this book because it fits the (micro) service world and the CD process.

Docker entails several additional technologies, as follows:

- **Docker Hub**: This is a registry for Docker images
- **Kubernetes**: This is a container orchestrator

> **Information**
>
> In the first edition of this book, Docker Compose and Docker Swarm were presented as tools for clustering and scheduling multi-container applications. Since that time, however, Kubernetes has become the market leader and is used instead.

Jenkins

Jenkins is by far the most popular automation server on the market. It helps create CI and CD pipelines and, in general, any other automated sequence of scripts. Highly plugin-oriented, it has a great community that constantly extends it with new features. What's more, it allows us to write the pipeline as code and supports distributed build environments.

Ansible

Ansible is an automation tool that helps with software provisioning, configuration management, and application deployment. It is trending faster than any other configuration management engine and will soon overtake its two main competitors: Chef and Puppet. It uses an agentless architecture and integrates smoothly with Docker.

GitHub

GitHub is the best of all hosted version control systems. It provides a very stable system, a great web-based UI, and a free service for public repositories. Having said that, any source control management service or tool will work with CD, irrespective of whether it's in the cloud or self-hosted, and whether it's based on Git, SVN, Mercurial, or any other tool.

Java/Spring Boot/Gradle

Java has been the most popular programming language for years. That's why it will be used for most of the code examples in this book. Together with Java, most companies develop with the Spring framework, so we used it to create a simple web service to explain some concepts. Gradle is used as a build tool. It's still less popular than Maven, but it's trending much faster. As always, any programming language, framework, or build tool can be exchanged and the CD process would stay the same, so don't worry if your technology stack is different.

The other tools

Cucumber was chosen arbitrarily as the acceptance testing framework. Other similar solutions are FitNesse and Jbehave. For the database migration process, we will use Flyway, but any other tool would do, such as Liquibase.

Creating a complete CD system

You can look at how this book is organized from two perspectives.

The first one is based on the steps of the automated deployment pipeline. Each chapter takes you closer to the complete CD process. If you look at the names of the chapters, some of them are even named like the pipeline phases:

- The CI pipeline

- Automated acceptance testing

- Configuration management with Ansible

The rest of the chapters provide an introduction, summary, or additional information that's complementary to the process.

There is also a second perspective to the content of this book. Each chapter describes one piece of the environment, which, in turn, is well prepared for the CD process. In other words, this book presents, step by step, technology by technology, how to build a complete system. To help you get a feel of what we plan to build throughout this book, let's have a look at how the system will evolve in each chapter.

> **Note**
> Don't worry if you don't understand the concepts and terminology at this point. We will be learning everything from scratch in the corresponding chapters.

Introducing Docker

In *Chapter 2*, *Introducing Docker*, we will start from the center of our system and build a working application that's been packaged as a Docker image. The output of this chapter is presented in the following diagram:

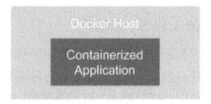

Figure 1.8 – Introducing Docker

A dockerized application (web service) is run as a container on a **Docker Host** and is reachable as it will run directly on the host machine. This is possible thanks to port forwarding (port publishing in Docker's terminology).

Configuring Jenkins

In *Chapter 3*, *Configuring Jenkins*, we will prepare the Jenkins environment. Thanks to the support of multiple agent (slave) nodes, it can handle the heavy concurrent load. The result is presented in the following diagram:

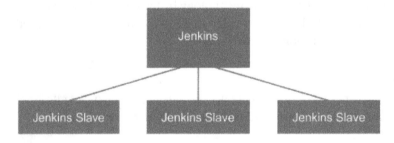

Figure 1.9 – Configuring Jenkins

The **Jenkins** master accepts a build request, but the execution is started at one of the **Jenkins Slave** (agent) machines. Such an approach provides horizontal scaling of the Jenkins environment.

The CI pipeline

In *Chapter 4, Continuous Integration Pipeline*, we'll show you how to create the first phase of the CD pipeline: the commit stage. The output of this chapter is shown in the following diagram:

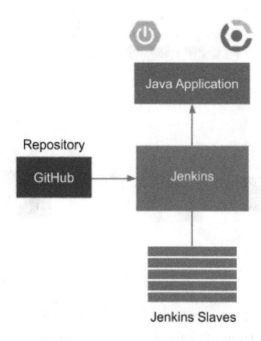

Figure 1.10 – The CI pipeline

The application is a simple web service written in Java with the Spring Boot framework. Gradle is used as a build tool and GitHub is used as the source code repository. Every commit to GitHub automatically triggers the Jenkins build, which uses Gradle to compile Java code, run unit tests, and perform additional checks (code coverage, static code analysis, and so on). Once the Jenkins build is complete, a notification is sent to the developers.

After this chapter, you will be able to create a complete CI pipeline.

Automated acceptance testing

In *Chapter 5, Automated Acceptance Testing*, we'll merge the two technologies mentioned in this book's title: *Docker* and *Jenkins*. This will result in the system presented in the following diagram:

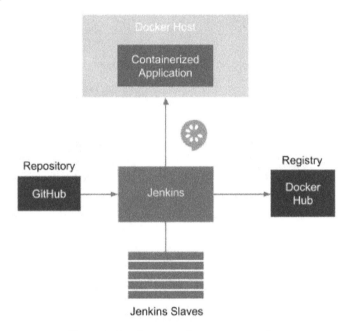

Figure 1.11 – Automated acceptance testing

The additional elements in the preceding diagram are related to the automated acceptance testing stage:

- **Docker Registry**: After the CI phase, the application is packaged into a JAR file and then as a Docker image. That image is then pushed to the **Docker Registry**, which acts as storage for dockerized applications.

- **Docker Host**: Before performing the acceptance test suite, the application must be started. Jenkins triggers a **Docker Host** machine to pull the dockerized application from the **Docker Registry** and starts it.

- **Cucumber**: After the application is started on the **Docker Host**, Jenkins runs a suite of acceptance tests written in the **Cucumber** framework.

Clustering with Kubernetes

In *Chapter 6, Clustering with Kubernetes*, we replace a single Docker host with a Kubernetes cluster and a single standalone application with two dependent containerized applications. The output is the environment shown in the following diagram:

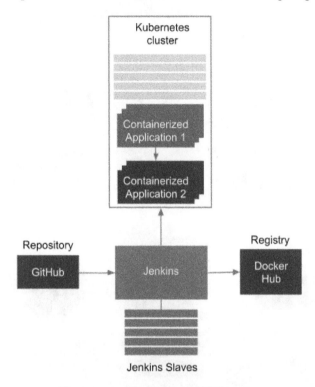

Figure 1.12 – Clustering with Kubernetes

Kubernetes provides an abstraction layer for a set of Docker hosts and allows simple communication between dependent applications. We no longer have to think about which machine our applications are deployed on. All we care about is the number of instances.

Configuration management with Ansible

In *Chapter 7, Configuration Management with Ansible*, we will create multiple environments using Ansible. The output is presented in the following diagram:

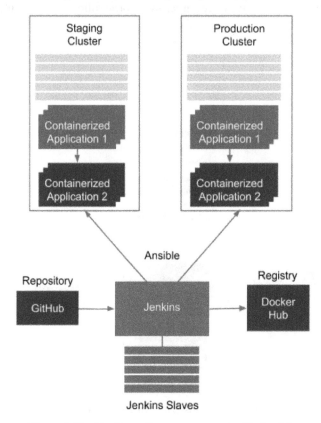

Figure 1.13 – Configuration management with Ansible

Ansible takes care of the environments and lets you deploy the same applications on multiple machines. As a result, we have a mirrored environment for testing and production.

In this chapter, we'll also touch on IaC and show you how to use Terraform if you use cloud environments.

The CD pipeline/advanced CD

In the last two chapters – that is, *Chapter 8, Continuous Delivery Pipeline*, and *Chapter 9, Advanced Continuous Delivery* – we will deploy the application to the staging environment, run the acceptance testing suite, and release the application to the production environment, usually in many instances. The final improvement is that we'll be able to automatically manage the database schemas using Flyway migrations that have been integrated into the delivery process. The final environment that will be created in this book is shown in the following diagram:

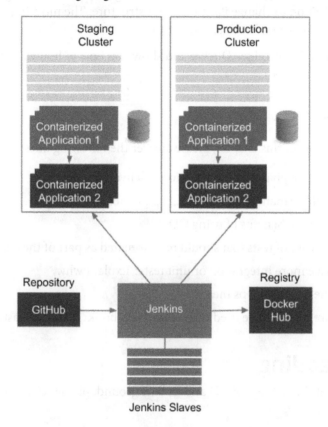

Figure 1.14 – The CD pipeline/advanced CD

I hope you are already excited by what we plan to build throughout this book. We will approach this step by step, explaining every detail and all the possible options to help you understand the procedures and tools. After reading this book, you will be able to introduce or improve the CD process in your projects.

Summary

In this chapter, we introduced the CD process, including the idea behind it, its prerequisites, and the tools that will be used throughout this book. The key takeaway from this chapter is that the delivery process that's currently used in most companies has significant shortcomings and can be improved using modern automation tools. The CD approach provides several benefits, of which the most significant ones are fast delivery, a fast feedback cycle, and low-risk releases. The CD pipeline consists of three stages: CI, automated acceptance testing, and configuration management. Introducing CD usually requires the organization to change its culture and structure. The most important tools in the context of CD are Docker, Jenkins, and Ansible.

In the next chapter, we'll introduce Docker and show you how to build a dockerized application.

Questions

To test your knowledge of this chapter, please answer the following questions:

1. What are the three phases of the traditional delivery process?
2. What are the three main stages of the CD pipeline?
3. Name at least three benefits of using CD.
4. What are the types of tests that should be automated as part of the CD pipeline?
5. Should we have more integration or unit tests? Explain why.
6. What does the term DevOps mean?
7. What software tools will be used throughout this book? Name at least four.

Further reading

To learn more about the concept of CD and its background, please refer to the following resources:

- *Continuous Delivery*, by Jez Humble and David Farley: `https://continuousdelivery.com/`
- *TestPyramid*, by Martin Fowler: `https://martinfowler.com/bliki/TestPyramid.html`
- *Succeeding with Agile: Software Development Using Scrum*, by Mike Cohn
- *Building Microservices: Designing Fine-Grained Systems*, by Sam Newman

2
Introducing Docker

In this chapter, we will discuss how the modern **continuous delivery (CD)** process looks by introducing Docker, the technology that changed the **information technology (IT)** industry and the way servers are used.

This chapter covers the following topics:

- What is Docker?
- Installing Docker
- Running Docker hello-world
- Docker components
- Docker applications
- Building Docker images
- Docker container states
- Docker networking
- Using Docker volumes
- Using names in Docker
- Docker cleanup
- Docker commands overview

Technical requirements

To complete this chapter, you'll need to meet the following hardware/software requirements:

- At least 4 **gigabytes (GB)** of **random-access memory (RAM)**

- macOS 10.15+, Windows 10/11 Pro 64-bit, Ubuntu 20.04+, or other Linux operating systems

All the examples and solutions to the exercises can be found at `https://github.com/PacktPublishing/Continuous-Delivery-With-Docker-and-Jenkins-3rd-Edition/tree/main/Chapter02`.

Code in Action videos for this chapter can be viewed at `https://bit.ly/3LJv1n6`.

What is Docker?

Docker is an open source project designed to help with application deployment using software containers. This approach means running applications together with the complete environment (files, code libraries, tools, and so on). Therefore, Docker—similar to virtualization—allows an application to be packaged into an image that can be run everywhere.

Containerization versus virtualization

Without Docker, isolation and other benefits can be achieved with the use of hardware virtualization, often called **virtual machines (VMs)**. The most popular solutions are VirtualBox, VMware, and parallels. A VM emulates a computer architecture and provides the functionality of a physical computer. We can achieve complete isolation of applications if each of them is delivered and run as a separate VM image.

The following diagram presents the concept of virtualization:

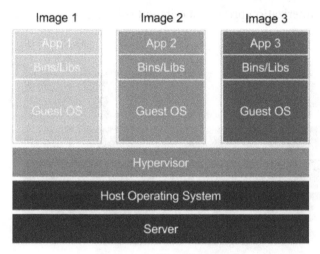

Figure 2.1 – Virtualization

Each application is launched as a separate image with all dependencies and a guest operating system. Images are run by the **hypervisor**, which emulates the physical computer architecture. This method of deployment is widely supported by many tools (such as Vagrant) and dedicated to development and testing environments. Virtualization, however, has three significant drawbacks, as outlined here:

- **Low performance**: The VM emulates the whole computer architecture to run the guest operating system, so there is a significant overhead associated with executing each operation.

- **High resource consumption**: Emulation requires a lot of resources and has to be done separately for each application. This is why, on a standard desktop machine, only a few applications can be run simultaneously.

- **Large image size**: Each application is delivered with a full operating system, so deployment on a server implies sending and storing a large amount of data.

The concept of containerization presents a different solution, as we can see here:

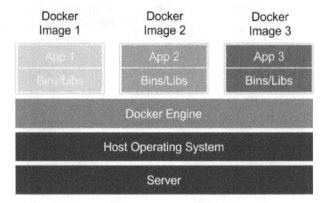

Figure 2.2 – Containerization

Each application is delivered together with its dependencies, but without the operating system. Applications interface directly with the host operating system, so there is no additional layer of the guest operating system. This results in better performance and no wasted resources. Moreover, shipped Docker images are significantly smaller.

Notice that, in the case of containerization, isolation happens at the level of the host operating system's processes. This doesn't mean, however, that the containers share their dependencies. Each of them has its own libraries in the right version, and if any of them is updated, it has no impact on the others. To achieve this, Docker Engine creates a set of Linux namespaces and control groups for the container. This is why Docker security is based on Linux kernel process isolation. This solution, although mature enough, could be considered slightly less secure than the complete operating system-based isolation offered by VMs.

The need for Docker

Docker containerization solves a number of problems seen in traditional software delivery. Let's take a closer look.

Environment

Installing and running software is complex. You need to make decisions about the operating system, resources, libraries, services, permissions, other software, and everything your application depends on. Then, you need to know how to install it. What's more, there may be some conflicting dependencies. *What do you do then? What if your software needs an upgrade of a library, but the other resources do not?* In some companies, such issues are solved by having *classes of applications*, and each class is served by a dedicated server, such as a server for web services with Java 7, and another one for batch jobs with Java 8. This solution, however, is not balanced in terms of resources and requires an army of IT operations teams to take care of all the production and test servers.

Another problem with the environment's complexity is that it often requires a specialist to run an application. A less technical person may have a hard time setting up MySQL, **Open Database Connectivity (ODBC)**, or any other slightly more sophisticated tool. This is particularly true for applications not delivered as an operating system-specific binary but that require source code compilation or any other environment-specific configuration.

Isolation

Keep the workspace tidy. One application can change the behavior of another one. Imagine what could happen. Applications share one filesystem, so if application *A* writes something to the wrong directory, application *B* reads the incorrect data. They share resources, so if there is a memory leak in application *A*, it can freeze not only itself but also application *B*. They share network interfaces, so if applications *A* and *B* both use port 8080, one of them will crash. Isolation concerns the security aspects, too. Running a buggy application or malicious software can cause damage to other applications. This is why it is a much safer approach to keep each application inside a separate sandbox, which limits the scope of possible damage to the application itself.

Organizing applications

Servers often end up looking messy, with a ton of running applications nobody knows anything about. *How will you check which applications are running on the server and which dependencies each of them is using?* They could depend on libraries, other applications, or tools. Without the exhaustive documentation, all we can do is look at the running processes and start guessing. Docker keeps things organized by having each application as a separate container that can be listed, searched, and monitored.

Portability

Write once, run anywhere, said the slogan while advertising the earliest versions of Java. Indeed, Java addresses the portability issue quite well. However, I can still think of a few cases where it fails; for example, the incompatible native dependencies or the older version of the Java Runtime. Moreover, not all software is written in Java.

Docker moves the concept of portability one level higher; if the Docker version is compatible, the shipped software works correctly, regardless of the programming language, operating system, or environment configuration. Docker, then, can be expressed by the following slogan: *Ship the entire environment instead of just code.*

Kittens and cattle

The difference between traditional software deployment and Docker-based deployment is often expressed with an analogy of kittens and cattle. Everybody likes kittens. Kittens are unique. Each has its own name and needs special treatment. Kittens are treated with emotion. We cry when they die. On the contrary, cattle exist only to satisfy our needs. Even the form *cattle* is singular since it's just a pack of animals treated together—no naming, no uniqueness. Surely, they are unique (the same as each server is unique), but this is irrelevant. This is why the most straightforward explanation of the idea behind Docker is *treat your servers like cattle, not pets.*

Alternative containerization technologies

Docker is not the only containerization system available on the market. Actually, the first versions of Docker were based on the open source **Linux Containers (LXC)** system, which is an alternative platform for containers. Other known solutions are **Windows Server containers**, **OpenVZ**, and **Linux Server**. Docker, however, overtook all other systems because of its simplicity, good marketing, and startup approach. It works under most operating systems, allows you to do something useful in less than 15 minutes, and has a lot of simple-to-use features, good tutorials, a great community, and probably the best logo in the IT industry!

We already understand the idea of Docker, so let's move on to the practical part and start from the beginning: Docker installation.

Installing Docker

Docker's installation process is quick and simple. Currently, it's supported on most Linux operating systems, and a wide range of them have dedicated binaries provided. macOS and Windows are also well supported with native applications. However, it's important to understand that Docker is internally based on the Linux kernel and its specifics, and this is why, in the case of macOS and Windows, it uses VMs (HyperKit for macOS and Hyper-V for Windows) to run the Docker Engine environment.

Prerequisites for Docker

The Docker Community Edition requirements are specific for each operating system, as outlined here:

- **macOS**:

    - macOS 10.15 or newer

    - At least 4 GB of RAM

    - No VirtualBox prior to version 4.3.30 installed

- **Windows**:

    - 64-bit Windows 10/11

    - The Hyper-V package enabled

    - At least 4 GB of RAM

- **Linux**:

    - 64-bit architecture

    - Linux kernel 3.10 or later

If your machine does not meet these requirements, the solution is to use **VirtualBox** with the Ubuntu operating system installed. This workaround, even though it sounds complicated, is not necessarily the worst method, especially considering that the Docker Engine environment is virtualized anyway in the case of macOS and Windows. Furthermore, Ubuntu is one of the best-supported systems for using Docker.

> Information
>
> All examples in this book have been tested on the Ubuntu 20.04 operating system.

Installing on a local machine

The Docker installation process is straightforward and is described in detail on its official page: `https://docs.docker.com/get-docker/`.

Docker Desktop

The simplest way to use Docker in your local environment is to install Docker Desktop. This way, in just a few minutes, you have a complete Docker development environment all set up and running. For Windows and macOS users, Docker Desktop provides a native application that hides all the setup difficulties behind the scenes. Technically, Docker Engine is installed inside a VM because Docker requires the Linux kernel to operate. Nevertheless, as a user, you don't even need to think about this—you install Docker Desktop and you are ready to start using the `docker` command. You can see an overview of Docker Desktop in the following screenshot:

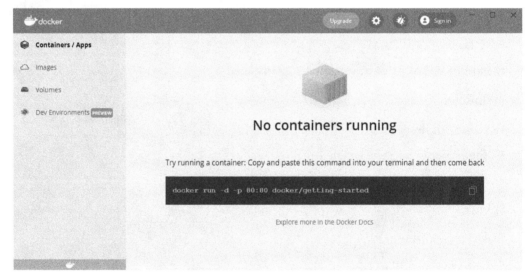

Figure 2.3 – Docker Desktop

Apart from Docker Engine, Docker Desktop provides a number of additional features, as follows:

- A **user interface** (**UI**) to display images, containers, and volumes
- A local Kubernetes cluster
- Automatic Docker updates
- Volume mounting with the local filesystem integration

- (Windows) Support for Windows containers

- (Windows) Integration with **Windows Subsystem for Linux (WSL)/WSL version 2 (WSL2)**

> **Note**
>
> Please visit `https://docs.docker.com/get-docker/` for Docker Desktop installation guides.

Docker for Ubuntu

Visit `https://docs.docker.com/engine/install/ubuntu/` to find a guide on how to install Docker on an Ubuntu machine.

In the case of Ubuntu 20.04, I've executed the following commands:

```
$ sudo apt-get update
$ sudo apt-get -y install ca-certificates curl gnupg
lsb-release
$ curl -fsSL https://download.docker.com/linux/ubuntu/gpg
| sudo gpg --dearmor -o /usr/share/keyrings/docker-archive-
keyring.gpg
$ echo "deb [arch=$(dpkg --print-architecture) signed-by=/usr/
share/keyrings/docker-archive-keyring.gpg] https://download.
docker.com/linux/ubuntu $(lsb_release -cs) stable" | sudo tee /
etc/apt/sources.list.d/docker.list > /dev/null
$ sudo apt-get update
$ sudo apt-get -y install docker-ce docker-ce-cli containerd.io
```

After all operations are completed, Docker should be installed. However, at the moment, the only user allowed to use Docker commands is `root`. This means that the `sudo` keyword must precede every Docker command.

We can enable other users to use Docker by adding them to the `docker` group, as follows:

```
$ sudo usermod -aG docker <username>
```

After a successful logout, everything is set up. With the latest command, however, we need to take some precautions not to give the Docker permissions to an unwanted user and thereby create a vulnerability in the Docker Engine environment. This is particularly important in the case of installation on the server machine.

Docker for other Linux distributions

Docker supports most Linux distributions and architectures. For details, please check the official page at `https://docs.docker.com/engine/install/`.

Testing the Docker installation

No matter which installation you've chosen (macOS, Windows, Ubuntu, Linux, or something else), Docker should be set up and ready. The best way to test it is to run the `docker info` command. The output message should be similar to the following:

```
$ docker info
Containers: 0
  Running: 0
   Paused: 0
  Stopped: 0
   Images: 0
...
```

Installing on a server

In order to use Docker over the network, it's possible to either take advantage of cloud platform providers or manually install Docker on a dedicated server.

In the first case, the Docker configuration differs from one platform to another, but it is always very well described in dedicated tutorials. Most cloud platforms enable Docker hosts to be created through user-friendly web interfaces or describe exact commands to execute on their servers.

The second case (installing Docker manually) does require a few words, however.

Dedicated server

Installing Docker manually on a server does not differ much from the local installation.

Two additional steps are required, which include setting the Docker daemon to listen on the network socket and setting security certificates. These steps are described in more detail here:

1. By default, due to security reasons, Docker runs through a non-networked Unix socket that only allows local communication. It's necessary to add listening on the chosen network interface socket so that external clients can connect. In the case of Ubuntu, the Docker daemon is configured by `systemd`, so, in order to change the configuration of how it's started, we need to modify one line in the `/lib/systemd/system/docker.service` file, as follows:

    ```
    ExecStart=/usr/bin/dockerd -H <server_ip>:2375
    ```

    By changing this line, we enabled access to the Docker daemon through the specified **Internet Protocol** (**IP**) address. All the details on the `systemd` configuration can be found at `https://docs.docker.com/config/daemon/systemd/`.

2. This step of server configuration concerns Docker security certificates. This enables only clients authenticated by a certificate to access the server. A comprehensive description of the Docker certificate configuration can be found at `https://docs.docker.com/engine/security/protect-access/`. This step isn't strictly required; however, unless your Docker daemon server is inside a firewalled network, it is essential.

    > **Information**
    >
    > If your Docker daemon is run inside a corporate network, you have to configure the **HyperText Transfer Protocol** (**HTTP**) proxy. A detailed description can be found at `https://docs.docker.com/config/daemon/systemd/`.

The Docker environment is set up and ready, so we can start the first example.

Running Docker hello-world

Enter the following command into your console:

```
$ docker run hello-world
Unable to find image 'hello-world:latest' locally
latest: Pulling from library/hello-world
1b930d010525: Pull complete
Digest: sha256:2557e3c07ed1e38f26e389462d03ed-
943586f744621577a99efb77324b0fe535
Status: Downloaded newer image for hello-world:latest


Hello from Docker!
This message shows that your installation appears to be working
correctly.

...
```

Congratulations! You've just run your first Docker container. I hope you can already see how simple Docker is. Let's examine what happened under the hood, as follows:

1. You ran the Docker client with the run command.

2. The Docker client contacted the Docker daemon and asked to create a container from the image called hello-world.

3. The Docker daemon checked whether it contained the hello-world image locally and, since it didn't, requested the hello-world image from the remote Docker Hub registry.

4. The Docker Hub registry contained the hello-world image, so it was pulled into the Docker daemon.

5. The Docker daemon created a new container from the hello-world image that started the executable producing the output.

6. The Docker daemon streamed this output to the Docker client.

7. The Docker client sent it to your Terminal.

The projected flow is represented in the following diagram:

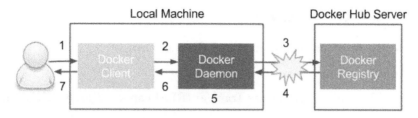

Figure 2.4 – Steps of the docker run command execution

Let's now look at each Docker component that was illustrated in this section.

Docker components

Docker is actually an ecosystem that includes a number of components. Let's describe all of them, starting with a closer look at the Docker client-server architecture.

Docker client and server

Let's look at the following diagram, which presents the Docker Engine architecture:

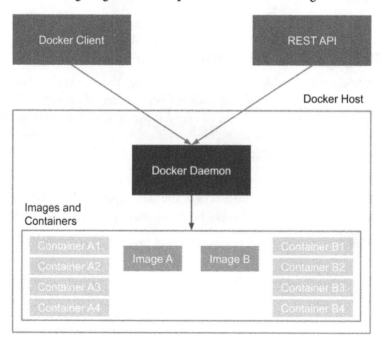

Figure 2.5 – Docker client-server architecture

Docker Engine consists of the following three components:

- A **Docker daemon** (server) running in the background
- A **Docker Client** running as a command tool
- A **Docker REpresentational State Transfer (REST) application programming interface (API)**

Installing Docker means installing all the components so that the Docker daemon runs on our computer all the time as a service. In the case of the `hello-world` example, we used the Docker client to interact with the Docker daemon; however, we could do exactly the same thing using the REST API. Also, in the case of the `hello-world` example, we connected to the local Docker daemon. However, we could use the same client to interact with the Docker daemon running on a remote machine.

> Tip
> To run the Docker container on a remote machine, you can use the `-H` option: `docker -H <server_ip>:2375 run hello-world`.

Docker images and containers

An **image** is a stateless building block in the Docker world. You can think of an image as a collection of all the files necessary to run your application, together with the recipe on how to run it. An image is stateless, so you can send it over the network, store it in the registry, name it, version it, and save it as a file. Images are layered, which means that you can build an image on top of another image.

A container is a running instance of an image. We can create many containers from the same image if we want to have many instances of the same application. Since containers are stateful, this means we can interact with them and make changes to their states.

Let's look at the following example of a **container** and **image** layered structure:

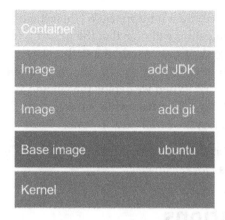

Figure 2.6 – Layered structure of Docker images

At the bottom, there is always the base image. In most cases, this represents an operating system, and we build our images on top of the existing base images. It's technically possible to create your own base images; however, this is rarely needed.

In our example, the `ubuntu` base image provides all the capabilities of the Ubuntu operating system. The `add git` image adds the Git toolkit. Then, there is an image that adds the **Java Development Kit** (**JDK**) environment. Finally, on the top, there is a container created from the `add JDK` image. Such a container is able, for example, to download a Java project from the GitHub repository and compile it to a **Java ARchive** (**JAR**) file. As a result, we can use this container to compile and run Java projects without installing any tools on our operating system.

It is important to note that layering is a very smart mechanism to save bandwidth and storage. Imagine that we have the following application that is also based on Ubuntu:

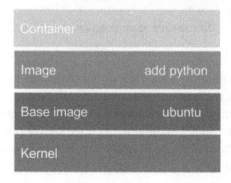

Figure 2.7 – Reusing layers of Docker images

This time, we files will use the Python interpreter. While installing the add python image, the Docker daemon will note that the ubuntu image is already installed, and what it needs to do is only to add the Python layer, which is very small. So, the ubuntu image is a dependency that is reused. The same applies if we would like to deploy our image in the network. When we deploy the Git and JDK application, we need to send the whole ubuntu image. However, while subsequently deploying the Python application, we need to send just the small add python layer.

Now that we understand what the Docker ecosystem consists of, let's describe how we can run applications packaged as Docker images.

Docker applications

A lot of applications are provided in the form of Docker images that can be downloaded from the internet. If we know the image name, it would be enough to run it in the same way we did with the hello-world example. *How can we find the desired application image on Docker Hub?* Let's take **MongoDB** as an example. These are the steps we need to follow:

1. If we want to find it on Docker Hub, we have two options, as follows:

    - Search on the Docker Hub **Explore** page (https://hub.docker.com/search/).

    - Use the docker search command.

    In the second case, we can perform the following operation:

    ```
    $ docker search mongo
    NAME     DESCRIPTION                    STARS    OFFICIAL
    AUTOMATED
    mongo    MongoDB document databases...  8293     [OK]
    ...
    ```

2. There are many interesting options. *How do we choose the best image?* Usually, the most appealing one is the one without any prefix, since it means that it's an official Docker Hub image and should therefore be stable and maintained. The images with prefixes are unofficial, usually maintained as open source projects. In our case, the best choice seems to be mongo, so in order to run the MongoDB server, we can run the following command:

    ```
    $ docker run mongo
    Unable to find image 'mongo:latest' locally
    latest: Pulling from library/mongo
    ```

```
7b722c1070cd: Pull complete
...
Digest: sha256:a7c1784c83536a3c686ec6f0a1c570ad-
5756b94a1183af88c07df82c5b64663c
{"t":{"$date":"2021-11-17T12:23:12.379+00:00"},"s":"I",
"c":"CONTROL",  "id":23285,   "ctx":"-","msg":"Automati-
cally disabling TLS 1.0, to force-enable TLS 1.0 specify
--sslDisabledProtocols 'none'"}
...
```

That's all we need to do. MongoDB has started. Running applications as Docker containers is that simple because we don't need to think of any dependencies; they are all delivered together with the image. Docker can be treated as a useful tool to run applications; however, the real power lies in building your own Docker images that wrap the programs together with the environment.

> **Information**
>
> On the Docker Hub service, you can find a lot of applications; they store millions of different images.

Building Docker images

In this section, we will see how to build Docker images using two different methods: the docker commit command and a Dockerfile automated build.

docker commit

Let's start with an example and prepare an image with the Git and JDK toolkits. We will use Ubuntu 20.04 as a base image. There is no need to create it; most base images are available in the Docker Hub registry. Proceed as follows:

1. Run a container from ubuntu:20.04 and connect it to its command line, like this:

   ```
   $ docker run -i -t ubuntu:20.04 /bin/bash
   ```

   We've pulled the ubuntu:20.04 image, run it as a container, and then called the /bin/bash command in an interactive way (-i flag). You should see the Terminal of the container. Since containers are stateful and writable, we can do anything we want in its Terminal.

2. Install the Git toolkit, as follows:

```
root@dee2cb192c6c:/# apt-get update
root@dee2cb192c6c:/# apt-get install -y git
```

3. Check whether the Git toolkit is installed by running the following command:

```
root@dee2cb192c6c:/# which git
/usr/bin/git
```

4. Exit the container, like this:

```
root@dee2cb192c6c:/# exit
```

5. Check what has changed in the container by comparing its unique container **identifier (ID)** to the ubuntu image, as follows:

```
$ docker diff dee2cb192c6c
```

The preceding command should print a list of all files changed in the container.

6. Commit the container to the image, like this:

```
$ docker commit dee2cb192c6c ubuntu_with_git
```

We've just created our first Docker image. Let's list all the images of our Docker host to see whether the image is present, as follows:

$ docker images				
REPOSITORY	TAG	IMAGE ID	CREATED	SIZE
ubuntu_with_git MB	latest	f3d674114fe2	About a minute ago	205
ubuntu MB	20.04	20bb25d32758	7 days ago	87.5
mongo MB	latest	4a3b93a299a7	10 days ago	394
hello-world kB	latest	fce289e99eb9	2 weeks ago	1.84

As expected, we see `hello-world`, `mongo` (installed before), `ubuntu` (the base image pulled from Docker Hub), and the freshly built `ubuntu_with_git` image. By the way, we can observe that the size of each image corresponds to what we've installed on the image.

Now, if we create a container from the image, it will have the Git tool installed, as illustrated in the following code snippet:

```
$ docker run -i -t ubuntu_with_git /bin/bash
root@3b0d1ff457d4:/# which git
/usr/bin/git
root@3b0d1ff457d4:/# exit
```

Dockerfile

Creating each Docker image manually with the `commit` command could be laborious, especially in the case of build automation and the CD process. Luckily, there is a built-in language to specify all the instructions that should be executed to build a Docker image.

Let's start with an example similar to the one with Git. This time, we will prepare an `ubuntu_with_python` image, as follows:

1. Create a new directory and a file called `Dockerfile` with the following content:

    ```
    FROM ubuntu:20.04
    RUN apt-get update && \
        apt-get install -y python
    ```

2. Run the following command to create an `ubuntu_with_python` image:

    ```
    $ docker build -t ubuntu_with_python .
    ```

3. Check that the image was created by running the following command:

    ```
    $ docker images
    ```

REPOSITORY	TAG	IMAGE ID	CREATED	SIZE
ubuntu_with_python	latest	d6e85f39f5b7	About a minute ago	147 MB
ubuntu_with_git_and_jdk	latest	8464dc10abbb	3 minutes ago	580 MB
ubuntu_with_git	latest	f3d674114fe2	9 minutes ago	205 MB

ubuntu	20.04	20bb25d32758	7 days ago 87.5 MB
mongo	latest	4a3b93a299a7	10 days ago 394 MB
hello-world	latest	fce289e99eb9	2 weeks ago 1.84 kB

We can now create a container from the image and check that the Python interpreter exists in exactly the same way we did after executing the `docker commit` command. Note that the `ubuntu` image is listed only once even though it's the base image for both `ubuntu_with_git` and `ubuntu_with_python`.

In this example, we used the first two Dockerfile instructions, as outlined here:

- `FROM` defines an image on top of which the new image will be built
- `RUN` specifies the commands to run inside the container.

The other widely used instructions are detailed as follows:.

- `COPY/ADD` copies a file or a directory into the filesystem of the image.
- `ENTRYPOINT` defines which application should be run in the executable container.

A complete guide of all Dockerfile instructions can be found on the official Docker page at `https://docs.docker.com/engine/reference/builder/`.

Complete Docker application

We already have all the information necessary to build a fully working application as a Docker image. As an example, we will prepare, step by step, a simple Python `hello-world` program. The steps are always the same, no matter which environment or programming language we use.

Writing the application

Create a new directory and, inside this directory, create a `hello.py` file with the following content:

```
print "Hello World from Python!"
```

Close the file. This is the source code of our application.

Preparing the environment

Our environment will be expressed in the Dockerfile. We need instructions to define the following:

- Which base image should be used
- How to install the Python interpreter
- How to include `hello.py` in the image
- How to start the application

In the same directory, create the Dockerfile, like this:

```
FROM ubuntu:20.04
RUN apt-get update && \
    apt-get install -y python
COPY hello.py .
ENTRYPOINT ["python", "hello.py"]
```

Building the image

Now, we can build the image exactly the same way we did before, as follows:

```
$ docker build -t hello_world_python .
```

Running the application

We run the application by running the container, like this:

```
$ docker run hello_world_python
```

You should see a friendly **Hello World from Python!** message. The most interesting thing in this example is that we are able to run the application written in Python without having the Python interpreter installed in our host system. This is possible because the application packed as an image has the environment already included.

> **Tip**
> An image with the Python interpreter already exists in the Docker Hub service, so in a real-life scenario, it would be enough to use it.

Environment variables

We've run our first homemade Docker application. However, *what if the execution of the application depends on some conditions?*

For example, in the case of the production server, we would like to print Hello to the logs, not to the console, or we may want to have different dependent services during the testing phase and the production phase. One solution would be to prepare a separate Dockerfile for each case; however, there is a better way: environment variables.

Let's change our hello-world application to print Hello World from <name_ passed_as_environment_variable> !. In order to do this, we need to proceed with the following steps:

1. Change the hello.py Python script to use the environment variable, as follows:

    ```
    import os
    print "Hello World from %s !" % os.environ['NAME']
    ```

2. Build the image, like this:

    ```
    $ docker build -t hello_world_python_name .
    ```

3. Run the container passing the environment variable, like this:

    ```
    $ docker run -e NAME=Rafal hello_world_python_name
    Hello World from Rafal !
    ```

4. Alternatively, we can define an environment variable value in Dockerfile, such as the following:

    ```
    ENV NAME Rafal
    ```

5. Run the container without specifying the -e option, as follows:

    ```
    $ docker build -t hello_world_python_name_default .
    $ docker run hello_world_python_name_default
    Hello World from Rafal !
    ```

Environment variables are especially useful when we need to have different versions of the Docker container depending on its purpose; for example, to have separate profiles for production and testing servers.

> **Information**
> If an environment variable is defined both in the Dockerfile and as a flag, then the flag takes precedence.

Docker container states

Every application we've run so far was supposed to do some work and stop—for example, we've printed `Hello from Docker!` and exited. There are, however, applications that should run continuously, such as services.

To run a container in the background, we can use the `-d` (`--detach`) option. Let's try it with the `ubuntu` image, as follows:

```
$ docker run -d -t ubuntu:20.04
```

This command started the Ubuntu container but did not attach the console to it. We can see that it's running by using the following command:

```
$ docker ps
CONTAINER ID     IMAGE           COMMAND          STATUS
PORTS
NAMES
95f29bfbaadc     ubuntu:20.04     "/bin/bash"     Up 5 seconds
kickass_stonebraker
```

This command prints all containers that are in a **running** state. *What about our old, already exited containers?* We can find them by printing all containers, like this:

```
$ docker ps -a
CONTAINER ID     IMAGE           COMMAND          STATUS          PORTS
NAMES
95f29bfbaadc     ubuntu:20.04     "/bin/bash"     Up 33 seconds
kickass_stonebraker
34080d914613     hello_world_python_name_default "python hello.
py" Exited lonely_newton
7ba49e8ee677 hello_world_python_name "python hello.py" Exited
mad_turing
dd5eb1ed81c3 hello_world_python "python hello.py" Exited
thirsty_bardeen
...
```

Note that all the old containers are in an **exited** state. There are two more states we haven't observed yet: **paused** and **restarting**.

All of the states and the transitions between them are presented in the following diagram:

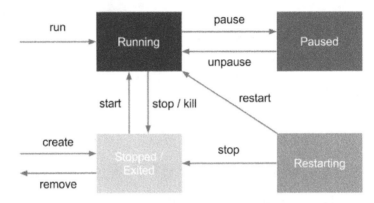

Figure 2.8 – Docker container states

Pausing Docker containers is very rare, and technically, it's done by freezing the processes using the SIGSTOP signal. Restarting is a temporary state when the container is run with the --restart option to define a restarting strategy (the Docker daemon is able to automatically restart the container in case of failure).

The preceding diagram also shows the Docker commands used to change the Docker container state from one state to another.

For example, we can stop running the Ubuntu container, as shown here:

```
$ docker stop 95f29bfbaadc
$ docker ps
CONTAINER ID IMAGE COMMAND CREATED STATUS PORTS NAMES
```

> **Information**
>
> We've always used the docker run command to create and start a container. However, it's possible to just create a container without starting it (with docker create).

Having grasped the details of Docker states, let's describe the networking basics within the Docker world.

Docker networking

Most applications these days do not run in isolation; they need to communicate with other systems over the network. If we want to run a website, web service, database, or cache server inside a Docker container, we need to first understand how to run a service and expose its port to other applications.

Running services

Let's start with a simple example and run a Tomcat server directly from Docker Hub, as follows:

```
$ docker run -d tomcat
```

Tomcat is a web application server whose UI can be accessed by port 8080. Therefore, if we installed Tomcat on our machine, we could browse it at http://localhost:8080. In our case, however, Tomcat is running inside the Docker container.

We started it the same way we did with the first Hello World example. We can see that it's running, as follows:

```
$ docker ps
CONTAINER ID IMAGE    COMMAND             STATUS           PORTS
NAMES
d51ad8634fac tomcat   "catalina.sh run"  Up About a minute 8080/
tcp jovial_kare
```

Since it's run as a daemon (with the -d option), we don't see the logs in the console right away. We can, however, access it by executing the following code:

```
$ docker logs d51ad8634fac
```

If there are no errors, we should see a lot of logs, which indicates that Tomcat has been started and is accessible through port 8080. We can try going to http://localhost:8080, but we won't be able to connect. This is because Tomcat has been started inside the container and we're trying to reach it from the outside. In other words, we can reach it only if we connect with the command to the console in the container and check it there. *How do we make running Tomcat accessible from outside?*

We need to start the container, specifying the port mapping with the -p (--publish) flag, as follows:

```
-p, --publish <host_port>:<container_port>
```

So, let's first stop the running container and start a new one, like this:

```
$ docker stop d51ad8634fac
$ docker run -d -p 8080:8080 tomcat
```

After waiting a few seconds, Tomcat should have started, and we should be able to open its page—http://localhost:8080.

The following screenshot illustrates how Docker container ports are published:

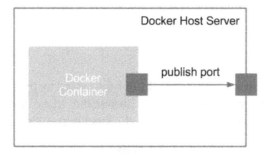

Figure 2.9 – Publishing Docker container ports

Such a simple port mapping command is sufficient in most common Docker use cases. We are able to deploy (micro) services as Docker containers and expose their ports to facilitate communication. However, let's dive a little deeper into what happened under the hood.

> **Information**
>
> Docker also allows us to publish to the specific host network interface with -p <ip>:<host_port>:<container_port>.

Container networks

We have connected to the application that is running inside the container. In fact, the connection is two-way because, if you remember from our previous examples, we executed the apt-get install commands from inside and the packages were downloaded from the internet. *How is this possible?*

If you check the network interfaces on your machine, you can see that one of the interfaces is called docker0, as illustrated here:

```
$ ifconfig docker0
docker0 Link encap:Ethernet HWaddr 02:42:db:d0:47:db
      inet addr:172.17.0.1 Bcast:0.0.0.0 Mask:255.255.0.0
...
```

The docker0 interface is created by the Docker daemon in order to connect with the Docker container. Now, we can see which interfaces are created inside the Tomcat Docker container created with the docker inspect command, as follows:

```
$ docker inspect 03d1e6dc4d9e
```

This prints all the information about the container configuration in **JavaScript Object Notation (JSON)** format. Among other things, we can find the part related to the network settings, as illustrated in the following code snippet:

```
"NetworkSettings": {
        "Bridge": "",
        "Ports": {
            "8080/tcp": [
                {
                    "HostIp": "0.0.0.0",
                    "HostPort": "8080"
                }
            ]
        },
        "Gateway": "172.17.0.1",
        "IPAddress": "172.17.0.2",
        "IPPrefixLen": 16,
}
```

> **Information**
>
> In order to filter the docker inspect response, we can use the --format option—for example, docker inspect --format '{{ .NetworkSettings.IPAddress }}' <container_id>.

We can observe that the Docker container has an IP address of 172.17.0.2 and it communicates with the Docker host with an IP address of 172.17.0.1. This means that in our previous example, we could access the Tomcat server even without the port forwarding, using http://172.17.0.2:8080. Nevertheless, in most cases, we run the Docker container on a server machine and want to expose it outside, so we need to use the -p option.

Note that, by default, the containers don't open any routes from external systems. We can change this default behavior by playing with the --network flag and setting it as follows:

- bridge (default): Network through the default Docker bridge
- none: No network
- container: Network joined with the other (specified) container
- host: Host's network stack
- NETWORK: User-created network (using the docker network create command)

The different networks can be listed and managed by the docker network command, as follows:

```
$ docker network ls
NETWORK ID      NAME      DRIVER    SCOPE
b3326cb44121    bridge    bridge    local
84136027df04    host      host      local
80c26af0351c    none      null      local
```

If we specify none as the network, we will not be able to connect to the container, and vice versa; the container has no network access to the external world. The host option makes the **container network interfaces** (**CNIs**) identical to the host. They share the same IP addresses, so everything started on the container is visible outside. The most popular option is the default one (bridge) because it lets us define explicitly which ports should be published and is both secure and accessible.

Exposing container ports

We mentioned a few times that the container exposes the port. In fact, if we dig deeper into the Tomcat image on GitHub (https://github.com/docker-library/tomcat), we can see the following line in the Dockerfile:

```
EXPOSE 8080
```

This Dockerfile instruction stipulates that port 8080 should be exposed from the container. However, as we have already seen, this doesn't mean that the port is automatically published. The EXPOSE instruction only informs users which ports they should publish.

Automatic port assignment

Let's try to run the second Tomcat container without stopping the first one, as follows:

```
$ docker run -d -p 8080:8080 tomcat
0835c95538aeca79e0305b5f19a5f96cb00c5d1c50bed87584cf-
ca8ec790f241
docker: Error response from daemon: driver failed program-
ming external connectivity on endpoint distracted_heyrovsky
(1b1cee9896ed99b9b804e4c944a3d9544adf72f1ef3f9c9f37b-
c985e9c30f452): Bind for 0.0.0.0:8080 failed: port is already
allocated.
```

This error may be common. In such cases, we have to either take care of the uniqueness of the ports on our own or let Docker assign the ports automatically using one of the following versions of the publish command:

- -p <container_port>: Publishes the container port to the unused host port

- -p (--publish-all): Publishes all ports exposed by the container to the unused host ports, as follows:

```
$ docker run -d -P tomcat
078e9d12a1c8724f8aa27510a6390473c1789aa49e7f8b-
14ddfaaa328c8f737b
$ docker port 078e9d12a1c8
8080/tcp -> 0.0.0.0:32772
```

We can see that the second Tomcat instance has been published to port 32772, so it can be browsed at http://localhost:32772.

After understanding Docker network basics, let's see how to provide a persistence layer for Docker containers using Docker volumes.

Using Docker volumes

Imagine that you would like to run a database as a container. You can start such a container and enter data. *Where is it stored? What happens when you stop the container or remove it?* You can start a new one, but the database will be empty again. Unless it's your testing environment, you'd expect to have your data persisted permanently.

A Docker volume is the Docker host's directory mounted inside the container. It allows the container to write to the host's filesystem as if it were writing to its own. The mechanism is presented in the following diagram:

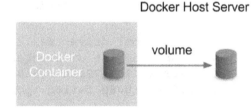

Figure 2.10 – Using a Docker volume

Docker volumes enable the persistence and sharing of a container's data. Volumes also clearly separate the processing from the data. Let's start with the following example:

1. Specify a volume with the -v <host_path>:<container_path> option and then connect to the container, as follows:

    ```
    $ docker run -i -t -v ~/docker_ubuntu:/host_directory
    ubuntu:20.04 /bin/bash
    ```

2. Create an empty file in host_directory in the container, like this:

    ```
    root@01bf73826624:/# touch /host_directory/file.txt
    ```

3. Check whether the file was created in the Docker host's filesystem by running the following command:

    ```
    root@01bf73826624:/# exit
    exit
    $ ls ~/docker_ubuntu/
    file.txt
    ```

4. We can see that the filesystem was shared and the data was therefore persisted permanently. Stop the container and run a new one to see if our file will still be there, as follows:

```
$ docker run -i -t -v ~/docker_ubuntu:/host_directory
ubuntu:20.04 /bin/bash
root@a9e0df194f1f:/# ls /host_directory/
file.txt
root@a9e0df194f1f:/# exit
```

5. Instead of specifying a volume with the -v flag, it's possible to specify it as an instruction in the Dockerfile, as in the following example:

```
VOLUME /host_directory
```

In this case, if we run the Docker container without the -v flag, the container's /host_directory path will be mapped into the host's default directory for volumes, /var/lib/docker/vfs/. This is a good solution if you deliver an application as an image and you know it requires permanent storage for some reason (for example, storing application logs).

> **Information**
>
> If a volume is defined both in a Dockerfile and as a flag, the flag command takes precedence.

Docker volumes can be much more complicated, especially in the case of databases. More complex use cases of Docker volumes are, however, outside the scope of this book.

> **Information**
>
> A very common approach to data management with Docker is to introduce an additional layer, in the form of data volume containers. A data volume container is a Docker container whose only purpose is to declare a volume. Then, other containers can use it (with the --volumes-from <container> option) instead of declaring the volume directly. Read more at https://docs.docker.com/storage/volumes/.

After understanding Docker volumes, let's see how we can use names to make working with Docker images/containers more convenient.

Using names in Docker

So far, when we've operated on containers, we've always used autogenerated names. This approach has some advantages, such as the names being unique (no naming conflicts) and automatic (no need to do anything). In many cases, however, it's better to give a user-friendly name to a container or an image.

Naming containers

There are two good reasons to name a container: convenience and the possibility of automation. Let's look at why, as follows:

- **Convenience**: It's simpler to make any operations on a container when addressing it by name than by checking the hashes or the autogenerated name.

- **Automation**: Sometimes, we would like to depend on the specific naming of a container.

For example, we would like to have containers that depend on each other and to have one linked to another. Therefore, we need to know their names.

To name a container, we use the --name parameter, as follows:

```
$ docker run -d --name tomcat tomcat
```

We can check (with docker ps) that the container has a meaningful name. Also, as a result, any operation can be performed using the container's name, as in the following example:

```
$ docker logs tomcat
```

Please note that when a container is named, it does not lose its identity. We can still address the container by its autogenerated hash ID, just as we did before.

Information
A container always has both an ID and a name. It can be addressed by either of them, and both are unique.

Tagging images

Images can be tagged. We already did this when creating our own images—for example, in the case of building the `hello_world_python` image, as illustrated here:

```
$ docker build -t hello_world_python .
```

The `-t` flag describes the tag of the image. If we don't use it, the image will be built without any tags and, as a result, we would have to address it by its ID (hash) in order to run the container.

The image can have multiple tags, and they should follow this naming convention:

```
<registry_address>/<image_name>:<version>
```

A tag consists of the following parts:

- `registry_address`: IP and port of the registry or the alias name
- `image_name`: Name of the image that is built—for example, `ubuntu`
- `version`: A version of the image in any form—for example, `20.04`, `20170310`

We will cover Docker registries in *Chapter 5, Automated Acceptance Testing*. If an image is kept on the official Docker Hub registry, we can skip the registry address. This is why we've run the `tomcat` image without any prefix. The last version is always tagged as the latest and it can also be skipped, so we've run the `tomcat` image without any suffix.

> **Information**
>
> Images usually have multiple tags; for example, all these three tags are the same image: `ubuntu:18.04`, `ubuntu:bionic-20190122`, and `ubuntu:bionic`.

Last but not least, we need to learn how to clean up after playing with Docker.

Docker cleanup

Throughout this chapter, we have created a number of containers and images. This is, however, only a small part of what you will see in real-life scenarios. Even when containers are not running, they need to be stored on the Docker host. This can quickly result in exceeding the storage space and stopping the machine. How can we approach this concern?

Cleaning up containers

First, let's look at the containers that are stored on our machine. Here are the steps we need to follow:

1. To print all the containers (irrespective of their state), we can use the docker ps -a command, as follows:

   ```
   $ docker ps -a
   CONTAINER ID IMAGE   COMMAND              STATUS    PORTS
   NAMES
   95c2d6c4424e tomcat "catalina.sh run" Up 5 minutes 8080/
   tcp tomcat
   a9e0df194f1f ubuntu:20.04 "/bin/bash" Exited
   jolly_archimedes
   01bf73826624 ubuntu:20.04 "/bin/bash" Exited
   suspicious_feynman
   ...
   ```

2. In order to delete a stopped container, we can use the docker rm command (if a container is running, we need to stop it first), as follows:

   ```
   $ docker rm 47ba1c0ba90e
   ```

3. If we want to remove all stopped containers, we can use the following command:

   ```
   $ docker container prune
   ```

4. We can also adopt a different approach and ask the container to remove itself as soon as it has stopped by using the ‑‑rm flag, as in the following example:

```
$ docker run --rm hello-world
```

In most real-life scenarios, we don't use stopped containers, and they are left only for debugging purposes.

Cleaning up images

Cleaning up images is just as important as cleaning up containers. They can occupy a lot of space, especially in the case of the CD process, when each build ends up in a new Docker image. This can quickly result in a *no space left on device* error. The steps are as outlined here:

1. To check all the images in the Docker container, we can use the docker images command, as follows:

```
$ docker images
REPOSITORY TAG                            IMAGE ID
CREATED        SIZE
hello_world_python_name_default latest 9a056ca92841 2
hours ago 202.6 MB
hello_world_python_name latest           72c8c50ffa89 2
hours ago 202.6 MB
...
```

2. To remove an image, we can call the following command:

```
$ docker rmi 48b5124b2768
```

3. In the case of images, the automatic cleanup process is slightly more complex. Images don't have states, so we cannot ask them to remove themselves when not used. A common strategy would be to set up a cron cleanup job, which removes all old and unused images. We could do this using the following command:

```
$ docker image prune
```

> **Information**
>
> If we have containers that use volumes, then, in addition to images and containers, it's worth thinking about cleaning up volumes. The easiest way to do this is to use the `docker volume prune` command.

With the cleaning up section, we've come to the end of the main Docker description. Now, let's do a short wrap-up and walk through the most important Docker commands.

> **Tip**
>
> Use the `docker system prune` command to remove all unused containers, images, and networks. Additionally, you can add the `-volumes` parameter to clean up volumes.

Docker commands overview

All Docker commands can be found by executing the following `help` command:

```
$ docker help
```

To see all the options of any particular Docker command, we can use `docker help <command>`, as in the following example:

```
$ docker help run
```

There is also a very good explanation of all Docker commands on the official Docker page at `https://docs.docker.com/engine/reference/commandline/docker/`. It's worth reading, or at least skimming, through.

In this chapter, we've covered the most useful commands and their options. As a quick reminder, let's walk through them, as follows:

Command	Explanation
`docker build`	Build an image from a Dockerfile.
`docker commit`	Create an image from a container.
`docker diff`	Show changes in a container.
`docker images`	List images.
`docker info`	Display Docker information.
`docker inspect`	Show the configuration of a Docker image/container.
`docker logs`	Show logs of a container.
`docker network`	Manage networks.
`docker port`	Show all ports exposed by a container.
`docker ps`	List containers.
`docker rm`	Remove a container.
`docker rmi`	Remove an image.
`docker run`	Run a container from an image.
`docker search`	Search for a Docker image in Docker Hub.
`docker start/stop/pause/unpause`	Manage a container's state.

Summary

In this chapter, we covered the Docker basics, which is enough for building images and running applications as containers. Here are the key takeaways.

The containerization technology addresses the issues of isolation and environment dependencies using Linux kernel features. This is based on a process separation mechanism, so, therefore, no real performance drop is observed. Docker can be installed on most systems but is supported natively only on Linux. Docker allows us to run applications from images available on the internet and to build our own images. An image is an application packed together with all the dependencies.

Docker provides two methods for building images—a Dockerfile or committing a container. In most cases, the first option is used. Docker containers can communicate over the network by publishing the ports they expose. Docker containers can share persistent storage using volumes. For the purpose of convenience, Docker containers should be named, and Docker images should be tagged. In the Docker world, there is a specific convention for how to tag images. Docker images and containers should be cleaned from time to time in order to save on server space and avoid *no space left on device* errors.

In the next chapter, we will look at the Jenkins configuration and find out how Jenkins can be used in conjunction with Docker.

Exercises

We've covered a lot of material in this chapter. To consolidate what we have learned, we recommend the following two exercises:

1. Run `CouchDB` as a Docker container and publish its port, as follows:

    > **Tip**
    >
    > You can use the `docker search` command to find the `CouchDB` image.

    I.   Run the container.

    II.   Publish the `CouchDB` port.

    III. Open the browser and check that `CouchDB` is available.

2. Create a Docker image with a REST service, replying `Hello World` to `localhost:8080/hello`. Use any language and framework you prefer. Here are the steps you need to follow:

    > **Tip**
    >
    > The easiest way to create a REST service is to use Python with the Flask framework (`https://flask.palletsprojects.com/`). Note that a lot of web frameworks, by default, start an application only on the localhost interface. In order to publish a port, it's necessary to start it on all interfaces (`app.run(host='0.0.0.0')` in the case of a Flask framework).

    I.   Create a web service application.

    II.   Create a Dockerfile to install dependencies and libraries.

    III. Build the image.

IV.   Run the container that is publishing the port.

V.   Check that it's running correctly by using the browser (or `curl`).

Questions

To verify the knowledge acquired from this chapter, please answer the following questionsUse L-numbering for this list

1.   What is the main difference between containerization (such as with Docker) and virtualization (such as with VirtualBox)?

2.   What are the benefits of providing an application as a Docker image? Name at least two.

3.   Can the Docker daemon be run natively on Windows and macOS?

4.   What is the difference between a Docker image and a Docker container?

5.   What does it mean when saying that Docker images have layers?

6.   What are two methods of creating a Docker image?

7.   Which command is used to create a Docker image from a Dockerfile?

8.   Which command is used to run a Docker container from a Docker image?

9.   In Docker terminology, what does it mean to publish a port?

10.   What is a Docker volume?

Further reading

If you're interested in getting a deeper understanding of Docker and related technologies, please have a look at the following resources:

- Docker documentation—*Get started*: `https://docs.docker.com/get-started/`

- *The Docker Book* by *James Turnbull*: `https://dockerbook.com/`

3

Configuring Jenkins

To start any continuous delivery process, we need an automation server such as Jenkins. However, configuring Jenkins can be difficult, especially when the amount of tasks assigned to it increases over time. What's more, since Docker allows the dynamic provisioning of Jenkins agents, is it worth spending time to configure everything correctly upfront, with scalability in mind?

In this chapter, we'll present Jenkins, which can be used separately or together with Docker. We will show that the combination of these two tools produces surprisingly good results – automated configuration and flexible scalability.

This chapter will cover the following topics:

- What is Jenkins?
- Installing Jenkins
- Jenkins – Hello World
- Jenkins architecture
- Configuring agents
- Custom Jenkins images
- Configuration and management

Technical requirements

To follow along with the instructions in this chapter, you'll need the following hardware/ software:

- Java 8+
- At least 4 GB of RAM
- At least 1 GB of free disk space
- Docker Engine installed

All the examples and solutions to the exercises in this chapter can be found on GitHub at `https://github.com/PacktPublishing/Continuous-Delivery-With-Docker-and-Jenkins-3rd-Edition/tree/main/Chapter03`.

Code in Action videos for this chapter can be viewed at `https://bit.ly/3DP02TW`.

What is Jenkins?

Jenkins is an open source automation server written in Java. With very active community-based support and a huge number of plugins, it is one of the most popular tools for implementing continuous integration and continuous delivery processes. Formerly known as **Hudson**, it was renamed after Oracle bought Hudson and decided to develop it as proprietary software. Jenkins was forked from Hudson but remained open source under the MIT license. It is highly valued for its simplicity, flexibility, and versatility.

Jenkins outshines other continuous integration tools and is the most widely used software of its kind. That's all possible because of its features and capabilities.

Let's walk through the most interesting parts of Jenkins' characteristics:

- **Language-agnostic**: Jenkins has a lot of plugins, which support most programming languages and frameworks. Moreover, since it can use any shell command and any software, it is suitable for every automation process imaginable.
- **Extensible by plugins**: Jenkins has a great community and a lot of available plugins (over a thousand). It also allows you to write your own plugins in order to customize Jenkins for your needs.
- **Portable**: Jenkins is written in Java, so it can be run on any operating system. For convenience, it is also delivered in a lot of versions – **Web Application Archive (WAR)** files, Docker images, Helm charts, Kubernetes operators, Windows binaries, macOS binaries, and Linux binaries.

- **Supports most Source Control Management (SCM) tools**: Jenkins integrates with virtually every source code management or build tool that exists. Again, because of its large community and number of plugins, there is no other continuous integration tool that supports so many external systems.

- **Distributed**: Jenkins has a built-in mechanism for the master/agent mode, which distributes its execution across multiple nodes, located on multiple machines. It can also use heterogeneous environments; for example, different nodes can have different operating systems installed.

- **Simplicity**: The installation and configuration process is simple. There is no need to configure any additional software or the database. It can be configured completely through a GUI, XML, or Groovy scripts.

- **Code-oriented**: Jenkins pipelines are defined as code. Also, Jenkins itself can be configured using YAML/XML files or Groovy scripts. That allows you to keep the configuration in the source code repository and helps in the automation of the Jenkins configuration.

Now that you have a basic understanding of Jenkins, let's move on to installing it.

Installing Jenkins

There are different methods of installing Jenkins, and you should choose the one that best suits your needs. Let's walk through all the options you have and then describe the most common choices in detail:

- **Servlet**: Jenkins is written in Java and natively distributed as a web application in the WAR format, dedicated to running inside an application server (such as Apache Tomcat or GlassFish); consider this option if you deploy all your applications as servlets.

- **Application**: The Jenkins WAR file embeds the Jetty application server, so it can be directly run with the Java command, and therefore, the **Java Runtime Environment (JRE)** is the only requirement to start Jenkins; consider this option if you use bare-metal servers and/or you need to install multiple Jenkins instances on one machine.

- **Dedicated package**: Jenkins is distributed for most operating systems in a form of dedicated packages (MSI for Windows, the Homebrew package for macOS, the deb package for Debian/Ubuntu, and so on); consider this option for the simplest installation and configuration if you use bare-metal servers.

- **Docker**: Jenkins is distributed in a form of a Docker image, and so the only requirement is to have Docker installed; consider this option for the simplest installation if you use Docker in your ecosystem.

- **Kubernetes**: Jenkins provides a Helm chart and a Kubernetes operator to simplify its installation, management, and scaling in a Kubernetes cluster; consider this option for the simplest Jenkins scaling and management.

- **Cloud**: Jenkins is hosted in a form of **Software as a Service (SaaS)** by a number of platforms; consider this option if you don't want to think about server maintenance and Jenkins installation.

Each installation method has its own pros and cons. Let's describe the most common approaches, starting from using a Jenkins Docker image.

> **Information**
>
> You can find a detailed description of each installation method at `https://www.jenkins.io/doc/book/installing/`.

Installing Jenkins with Docker

The Jenkins image is available in the Docker Hub registry, so in order to install its latest version, we should execute the following command:

```
$ docker run -p <host_port>:8080 -v <host_volume>:/var/jenkins_
home jenkins/jenkins
```

We need to specify the following parameters:

- The first `host_port` parameter: The port on which Jenkins is visible outside of the container.

- A second `host_volume` parameter: This specifies the directory where the Jenkins home is mapped. It needs to be specified as volume; therefore, it is persisted permanently because it contains the configuration, pipeline builds, and logs.

As an example, let's follow the installation steps:

1. **Prepare the volume directory**: We need a separate directory to keep the Jenkins data. Let's prepare one with the following commands:

   ```
   $ mkdir $HOME/jenkins_home
   ```

2. **Run the Jenkins container**: Let's run the container as a daemon and give it a proper name with the following command:

   ```
   $ docker run -d -p 8080:8080 \
     -v $HOME/jenkins_home:/var/jenkins_home \
     --name jenkins jenkins/jenkins
   ```

3. **Check whether Jenkins is running**: After a moment, we can check whether Jenkins has started correctly by printing the logs:

   ```
   $ docker logs jenkins
   Running from: /usr/share/jenkins/jenkins.war
   webroot: EnvVars.masterEnvVars.get("JENKINS_HOME")
   ...
   ```

> **Information**
>
> In the production environment, you may also want to set up some additional parameters; for details, please refer to https://www.jenkins.io/doc/book/installing/docker/.

After performing these steps, you can access your Jenkins instance at http://localhost:8080/.

Installing Jenkins with dedicated packages

If you don't use Docker on your servers, then the simplest way is to use dedicated packages. Jenkins supports most operating systems – for example, MSI for Windows, the Homebrew package for macOS, and the deb package for Debian/Ubuntu.

As an example, in the case of Ubuntu, it's enough to run the following commands to install Jenkins (and the required Java dependency):

```
$ sudo apt-get update
$ sudo apt-get -y install default-jdk
$ wget -q -O - https://pkg.jenkins.io/debian/jenkins.io.key |
sudo apt-key add -
$ sudo sh -c 'echo deb http://pkg.jenkins.io/debian-stable
binary/ > /etc/apt/sources.list.d/jenkins.list'
$ sudo apt-get update
$ sudo apt-get -y install jenkins
```

After successful installation, the Jenkins instance is accessible via `http://localhost:8080/`.

Initial configuration

No matter which installation you choose, starting Jenkins requires a few configuration steps. Let's walk through them step by step:

1. Open Jenkins in the browser, at `http://localhost:8080`.
2. Jenkins will ask for the administrator password. It can be found in the Jenkins logs:

```
$ docker logs jenkins
...
Jenkins initial setup is required. An admin user has been
created
and a password generated.
Please use the following password to proceed to
installation:

c50508effc6843a1a7b06f6491ed0ca6

...
```

3. After accepting the initial password, Jenkins asks whether to install the suggested plugins, which are adjusted for the most common use cases. Your answer depends on your needs, of course. However, as the first Jenkins installation, it's reasonable to let Jenkins install all the recommended plugins.

4. After the plugin installation, Jenkins asks you to set up a username, password, and other basic information. If you skip it, the token from *step 2* will be used as the admin password.

The installation is then complete, and you should see the **Jenkins** dashboard:

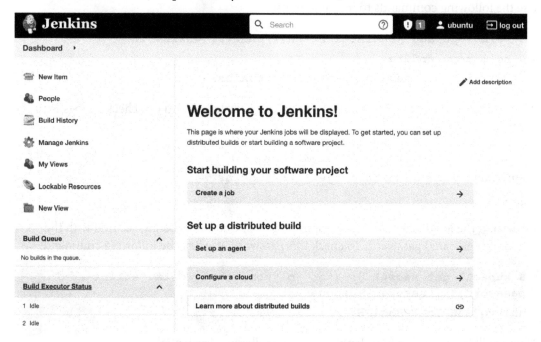

Figure 3.1 – A successful Jenkins installation

Now, let's see how to install Jenkins if your deployment environment is a Kubernetes cluster.

Installing Jenkins in Kubernetes

There are two methods of installing Jenkins in Kubernetes – a Helm chart and a Kubernetes operator. Let's look at the simpler option and use the Helm tool.

> **Tip**
>
> For more details about the Helm tool and its installation procedure, please visit `https://helm.sh/`.

Use the following commands to install Jenkins:

```
$ helm repo add jenkinsci https://charts.jenkins.io
$ helm repo update
$ helm install jenkins jenkinsci/jenkins
```

After executing the preceding commands, Jenkins is installed. You can check its logs with the following command:

```
$ kubectl logs sts/jenkins jenkins
Running from: /usr/share/jenkins/jenkins.war
...
```

By default, the Jenkins instance is configured with one admin account, secured with the randomly generated password. To check this password, execute the following command:

```
$ kubectl get secret jenkins -o jsonpath="{.data.jenkins-admin-
password}" | base64 --decode
nn1Pvq7asHPYz7EUHhc4PH
```

Now, you'll be able to log in to Jenkins with the following credentials:

- **Username**: admin
- **Password**: nn1Pvq7asHPYz7EUHhc4PH

By default, Jenkins is not exposed outside the Kubernetes cluster. To make it accessible from your local machine, run the following command:

```
$ kubectl port-forward sts/jenkins 8080:8080
```

After this, you can open your browser at `http://localhost:8080/` and log in with the aforementioned credentials.

> **Information**
>
> Please visit `https://www.jenkins.io/doc/book/installing/kubernetes/` for more information about installing Jenkins in Kubernetes.

One of the biggest benefits of installing Jenkins in the Kubernetes cluster instead of a single machine is that it provides horizontal scaling out of the box. Jenkins agents are automatically provisioned using Jenkins' Kubernetes plugin.

We will cover scaling Jenkins in the *Jenkins architecture* section and more about Kubernetes in *Chapter 6, Clustering with Kubernetes*. Now, let's see how you can use Jenkins in the cloud.

Jenkins in the cloud

If you don't want to install Jenkins yourself, there are companies that offer Jenkins hosted in the cloud. Note, however, that Jenkins was never built with a cloud-first approach in mind, so most offerings are, in fact, generic cloud solutions that help in installing and managing the Jenkins application for you.

The solution I recommend is Google Cloud Marketplace, which automatically deploys Jenkins in Google Kubernetes Engine. Read more at `https://cloud.google.com/jenkins`. Other companies that offer hosted Jenkins include Kamatera and Servana.

When we finally have Jenkins up and running, we are ready to create our first Jenkins pipeline.

Jenkins – Hello World

Everything in the entire IT world starts with the `Hello World` example, to show that the basics work fine. Let's follow this rule and use it to create the first Jenkins pipeline:

1. Click on **New Item**:

Figure 3.2 – New Item in the Jenkins web interface

2. Enter `hello world` as the item name, choose **Pipeline**, and click on **OK**:

Enter an item name

    hello world

» *Required field*

Freestyle project
This is the central feature of Jenkins. Jenkins will build your project, combining any SCM with any build system, and this can be even used for something other than software build.

Pipeline
Orchestrates long-running activities that can span multiple build agents. Suitable for building pipelines (formerly known as workflows) and/or organizing complex activities that do not easily fit in free-style job type.

Multi-configuration project
Suitable for projects that need a large number of different configurations, such as testing on multiple environments, platform-specific builds, etc.

    OK

Figure 3.3 – A new pipeline in the Jenkins web interface

3. There are a lot of options. We will skip them for now and go directly to the **Pipeline** section:

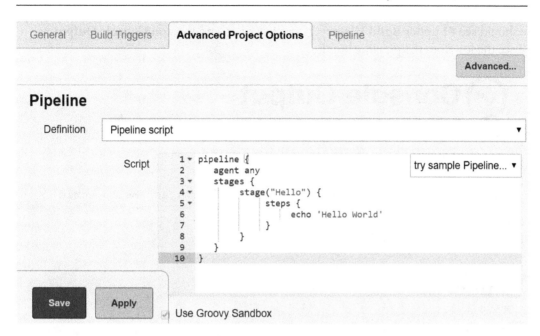

Figure 3.4 – Pipeline script in the Jenkins web interface

4. Then, in the **Script** textbox, we can enter the pipeline script:

```
pipeline {
    agent any
    stages {
        stage("Hello") {
            steps {
                echo 'Hello World'
            }
        }
    }
}
```

5. Click on **Save**.

6. Click on **Build Now**:

Figure 3.5 – Build Now in the Jenkins web interface

We should see **#1** under **Build History**. If we click on it, and then on **Console Output**, we will see the log from the pipeline build:

```
Started by user ubuntu
[Pipeline] Start of Pipeline
[Pipeline] node
Running on Jenkins in /var/jenkins_home/workspace/hello world
[Pipeline] {
[Pipeline] stage
[Pipeline] { (Hello)
[Pipeline] echo
Hello World
[Pipeline] }
[Pipeline] // stage
[Pipeline] }
[Pipeline] // node
[Pipeline] End of Pipeline
Finished: SUCCESS
```

Figure 3.6 – Console Output in the Jenkins web interface

The successful output in this first example means that Jenkins is installed correctly. Now, let's look at the possible Jenkins architecture.

> **Information**
> We will describe more about the pipeline syntax in *Chapter 4, Continuous Integration Pipeline*.

Jenkins architecture

Hello World is executed in almost no time at all. However, the pipelines are usually more complex, and time is spent on tasks such as downloading files from the internet, compiling source code, or running tests. One build can take from minutes to hours.

In common scenarios, there are also many concurrent pipelines. Usually, a whole team, or even a whole organization, uses the same Jenkins instance. *How can we ensure that the builds will run quickly and smoothly?*

Master and agents

Jenkins becomes overloaded sooner than it seems. Even in the case of a small (micro) service, the build can take a few minutes. That means that one team committing frequently can easily kill the Jenkins instance.

For that reason, unless the project is really small, Jenkins should not execute builds at all but delegate them to the agent (slave) instances. To be precise, the Jenkins server we're currently running is called the **Jenkins master**, and it can delegate execution tasks to **Jenkins agents**.

Let's look at a diagram presenting the master-agent interaction:

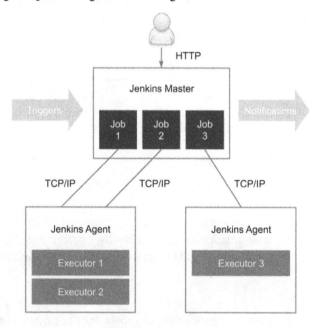

Figure 3.7 – The Jenkins master-agent interaction

In a distributed build environment, the Jenkins master is responsible for the following:

- Receiving build triggers (for example, after a commit to GitHub)
- Sending notifications (for example, email or Slack messages sent after a build failure)
- Handling HTTP requests (interaction with clients)
- Managing the build environment (orchestrating the job executions on agents)

The build agent is a machine that takes care of everything that happens after the build has started.

Since the responsibilities of the master and the agents are different, they have different environmental requirements:

- **Master**: This is usually (unless the project is really small) a dedicated machine with RAM ranging from 200 MB for small projects to 70+ GB for huge single-master projects.

- **Agent**: There are no general requirements (other than the fact that it should be capable of executing a single build; for example, if the project is a huge monolith that requires 100 GB of RAM, then the agent machine needs to satisfy these needs).

Agents should also be as generic as possible. For instance, if we have different projects – one in Java, one in Python, and one in Ruby – then it would be perfect if each agent could build any of these projects. In such a case, the agents can be interchanged, which helps to optimize the usage of resources.

> **Tip**
> If agents cannot be generic enough to match all projects, then it's possible to label (tag) agents and projects so that the given build will be executed on a given type of agent.

Scalability

As with everything in the software world, with growing usage, a Jenkins instance can quickly become overloaded and unresponsive. That is why we need to think upfront about scaling it up. There are two possible methods – **vertical scaling** and **horizontal scaling**.

Vertical scaling

Vertical scaling means that when the master's load grows, more resources are applied to the master's machine. So, when new projects appear in our organization, we buy more RAM, add CPU cores, and extend the HDD drives. This may sound like a no-go solution; however, it is used often, even by well-known organizations. Having a single Jenkins master set on ultra-efficient hardware has one very strong advantage – maintenance. Any upgrades, scripts, security settings, role assignments, or plugin installations have to be done in one place only.

Horizontal scaling

Horizontal scaling means that when an organization grows, more master instances are launched. This requires a smart allocation of instances to teams, and in extreme cases, each team can have its own Jenkins master. In that case, it might even happen that no agents are needed.

The drawbacks are that it may be difficult to automate cross-project integrations and that a part of the team's development time is spent on the maintenance of Jenkins. However, horizontal scaling has some significant advantages:

- Master machines don't need to be special, in terms of hardware.
- Different teams can have different Jenkins settings (for example, different sets of plugins).
- Teams usually feel better and work with Jenkins more efficiently if the instance is their own.
- If one master instance is down, it does not impact the whole organization.
- The infrastructure can be segregated into standard and mission-critical.

Test and production instances

Apart from the scaling approach, there is one more issue – *how to test the Jenkins upgrades, new plugins, or pipeline definitions.* Jenkins is critical to the whole company. It guarantees the quality of the software and, in the case of continuous delivery, deploys to the production servers. That is why it needs to be highly available, and it is definitely not for the purpose of testing. It means there should always be two instances of the same Jenkins infrastructure – **test** and **production**.

Sample architecture

We already know that there should be agents and (possibly multiple) masters and that everything should be duplicated in the test and production environments. However, *what would the complete picture look like?*

Luckily, there are a lot of companies that have published how they used Jenkins and what kind of architectures they created. It would be difficult to measure whether more of them preferred vertical or horizontal scaling, but it ranged from having only one master instance to having one master for each team.

Let's look at the example of Netflix to get a picture of a complete Jenkins infrastructure (Netflix shared it as a **planned infrastructure** at the Jenkins User Conference in San Francisco in 2012):

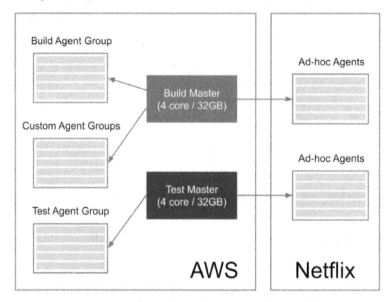

Figure 3.8 – The Jenkins infrastructure from Netflix

They have test and production master instances, with each of them owning a pool of agents and additional ad hoc agents. Altogether, it serves around 2,000 builds per day. Also, note that a part of their infrastructure is hosted on AWS and another part is on their own servers.

You should already have a rough idea of what the Jenkins infrastructure can look like, depending on the type of organization.

Now, let's focus on the practical aspects of setting the agents.

Configuring agents

You've seen what the agents are and when they can be used. However, *how do we set up an agent and let it communicate with the master?* Let's start with the second part of the question and describe the communication protocols between the master and the agent.

Communication protocols

In order for the master and the agent to communicate, a bidirectional connection has to be established.

There are different options for how it can be initiated:

- **SSH**: The master connects to the agent using the standard SSH protocol. Jenkins has an SSH client built in, so the only requirement is the **SSH daemon (sshd)** server configured on the agents. This is the most convenient and stable method because it uses standard Unix mechanisms.

- **Java web start**: A Java application is started on each agent machine, and the TCP connection is established between the Jenkins agent application and the master Java application. This method is often used if the agents are inside the firewalled network and the master cannot initiate the connection.

Once we know the communication protocols, let's look at how we can use them to set the agents.

Setting agents

At a low level, agents always communicate with the Jenkins master using one of the protocols described previously. However, at a higher level, we can attach agents to the master in various ways. The differences concern two aspects:

- **Static versus dynamic**: The simplest option is to add agents permanently in the Jenkins master. The drawback of such a solution is that we always need to manually change something if we need more (or fewer) agent nodes. A better option is to dynamically provision agents as they are needed.

- **Specific versus general-purpose**: Agents can be specific (for example, different agents for the projects based on Java 8 and Java 11) or general-purpose (an agent acts as a Docker host and a pipeline is built inside a Docker container).

These differences resulted in four common strategies for how agents are configured:

- Permanent agents
- Permanent Docker host agents
- Jenkins Swarm agents
- Dynamically provisioned Docker agents
- Dynamically provisioned Kubernetes agents

Let's examine each of the solutions.

Permanent agents

We will start with the simplest option, which is to permanently add specific agent nodes. It can be done entirely via the Jenkins web interface.

Configuring permanent agents

In the Jenkins master, when we open **Manage Jenkins** and then **Manage Nodes and Clouds**, we can view all the attached agents. Then, by clicking on **New Node**, giving it a name, setting its type to **Permanent Agent**, and confirming with the **Create** button, we should finally see the agent's setup page:

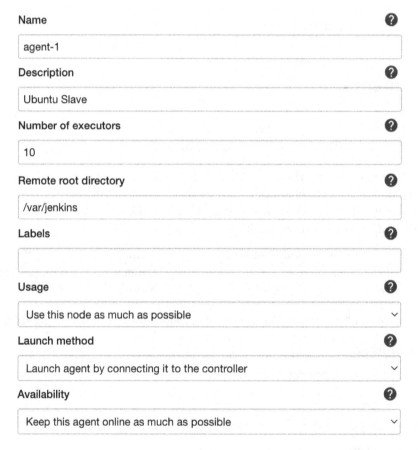

Figure 3.9 – Permanent agent configuration

Let's walk through the parameters we need to fill in:

- **Name**: This is the unique name of the agent.

- **Description**: This is a human-readable description of the agent.

- **Number of executors**: This is the number of concurrent builds that can be run on the agent.

- **Remote root directory**: This is the dedicated directory on the agent machine that the agent can use to run build jobs (for example, `/var/jenkins`); the most important data is transferred back to the master, so the directory is not critical.

- **Labels**: This includes the tags to match the specific builds (tagged the same) – for example, only projects based on Java 8.

- **Usage**: This is the option to decide whether the agent should only be used for matched labels (for example, only for acceptance testing builds), or for any builds.

- **Launch method**: This includes the following:

  - **Launch agent by connecting it to the controller**: Here, the connection will be established by the agent; it is possible to download the JAR file and the instructions on how to run it on the agent machine.

  - **Launch agent via execution of command on the controller**: This is the custom command run on the master to start the agent; in most cases, it will send the Java Web Start JAR application and start it on the agent (for example, `ssh <agent_hostname> java -jar ~/bin/slave.jar`).

  - **Launch agents via SSH**: Here, the master will connect to the agent using the SSH protocol.

- **Availability**: This is the option to decide whether the agent should be up all the time or the master should turn it offline under certain conditions.

> **Tip**
>
> The Java Web Start agent uses port `50000` for communication with the Jenkins master; therefore, if you use the Docker-based Jenkins master, you need to publish that port (`-p 50000:50000`).

When the agents are set up correctly, it's possible to update the master built-in node configuration with **Number of executors** set to 0 so that no builds will be executed on it, and it will only serve as the Jenkins UI and the build's coordinator.

> **Information**
>
> For more details and step-by-step instructions on how to configure permanent Jenkins agents, visit `https://www.jenkins.io/doc/book/using/using-agents/`.

Understanding permanent agents

As we've already mentioned, the drawback of such a solution is that we need to maintain multiple agent types (labels) for different project types. Such a situation is presented in the following diagram:

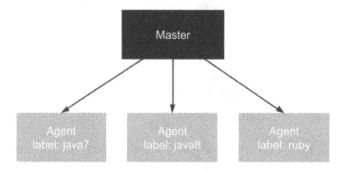

Figure 3.10 – Permanent agents

In our example, if we have three types of projects (**java7**, **java8**, and **ruby**), then we need to maintain three separately labeled (sets of) agents. That is the same issue we had while maintaining multiple production server types, as described in *Chapter 2, Introducing Docker*. We addressed the issue by having Docker Engine installed on the production servers. Let's try to do the same with Jenkins agents.

Permanent Docker host agents

The idea behind this solution is to permanently add general-purpose agents. Each agent is identically configured (with Docker Engine installed), and each build is defined along with the Docker image, inside of which the build is run.

Configuring permanent Docker host agents

The configuration is static, so it's done exactly the same way as we did with the permanent agents. The only difference is that we need to install Docker on each machine that will be used as an agent. Then, we usually don't need labels because all the agents can be the same. After the agents are configured, we define the Docker image in each pipeline script:

```
pipeline {
    agent {
        docker {
            image 'openjdk:8-jdk-alpine'
        }
    }
```

```
    . . .
}
```

When the build is started, the Jenkins agent starts a container from the Docker image, `openjdk:8-jdk-alpine`, and then executes all the pipeline steps inside that container. This way, we always know the execution environment and don't have to configure each agent separately, depending on the particular project type.

Understanding permanent Docker host agents

Looking at the same scenario we used for the permanent agents, the diagram looks like this:

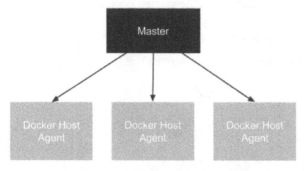

Figure 3.11 – Permanent Docker host agents

Each agent is exactly the same, and if we want to build a project that depends on Java 8, then we would define the appropriate Docker image in the pipeline script (instead of specifying the agent label).

Jenkins Swarm agents

So far, we have always had to permanently define each of the agents in the Jenkins master. Such a solution, although good enough in many cases, can be a burden if we need to frequently scale the number of agent machines. Jenkins Swarm allows you to dynamically add agents without the need to configure them in the Jenkins master.

Configuring Jenkins Swarm agents

The first step to using Jenkins Swarm is to install the **Swarm** plugin in Jenkins. We can do it through the Jenkins web UI, under **Manage Jenkins** and **Manage Plugins**. After this step, the Jenkins master is prepared for Jenkins agents to be dynamically attached.

The second step is to run the Jenkins Swarm agent application on every machine that will act as a Jenkins agent. We can do it using the `swarm-client.jar` application.

> **Information**
>
> The `swarm-client.jar` application can be downloaded from the Jenkins Swarm plugin page at `https://plugins.jenkins.io/swarm/`. On that page, you can also find all the possible options for its execution.

To attach the Jenkins Swarm agent node, run the following command:

```
$ java -jar path/to/swarm-client.jar -url ${JENKINS_URL}
-username ${USERNAME}
```

After successful execution, we should notice that a new agent has appeared on the Jenkins master, and when we run a build, it will be started on this agent.

Understanding Jenkins Swarm agents

Let's look at the following diagram that presents the Jenkins Swarm configuration:

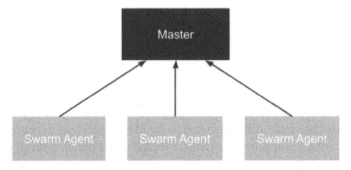

Figure 3.12 – Jenkins Swarm agents

Jenkins Swarm allows you to dynamically add agents, but it says nothing about whether to use specific or Docker-based agents, so we can use it for both. At first glance, Jenkins Swarm may not seem very useful. After all, we have moved setting agents from the master to the agent, but we still have to do it manually. However, with the use of a clustering system such as Kubernetes or Docker Swarm, Jenkins Swarm apparently enables the dynamic scaling of agents on a cluster of servers.

Dynamically provisioned Docker agents

Another option is to set up Jenkins to dynamically create a new agent each time a build is started. Such a solution is obviously the most flexible one, since the number of agents dynamically adjusts to the number of builds. Let's take a look at how to configure Jenkins this way.

Configuring dynamically provisioned Docker agents

First, we need to install the **Docker** plugin. As always, with Jenkins plugins, we can do this in **Manage Jenkins** and **Manage Plugins**. After the plugin is installed, we can start the following configuration steps:

1. Open the **Manage Jenkins** page.

2. Click on the **Manage Nodes and Clouds** link.

3. Click on the **Configure Clouds** link.

4. Click on **Add a new cloud** and choose **Docker**.

5. Fill in the Docker agent details, as shown in the following screenshot:

Figure 3.13 – Docker agent configuration

6. Most parameters do not need to be changed; however (apart from selecting **Enabled**), we need to at least set the Docker host URL (the address of the Docker host machine where agents will be run).

> **Tip**
> If you plan to use the same Docker host where Jenkins is running, then the Docker daemon needs to listen on the `docker0` network interface. You can do it in a similar way as to what's described in the *Installing on a server* section in *Chapter 2, Introducing Docker*, by changing one line in the `/lib/systemd/system/docker.service` file to `ExecStart=/usr/bin/dockerd -H 0.0.0.0:2375 -H fd://`.

7. Click on **Docker Agent templates...** and select **Add Docker Template**.

8. Fill in the details about the Docker agent image:

Figure 3.14 – Docker Agent templates configuration

We can use the following parameters:

- **Docker Image**: The most popular agent image from the Jenkins community is jenkins/agent (used for the default connect method, which is **Attach Docker container**).

- **Instance Capacity**: This defines the maximum number of agents running at the same time; to start with, it can be set to 10.

> **Information**
>
> Instead of jenkins/agent, it's possible to build and use your own agent images. This may be helpful in the case of specific environment requirements – for example, you need Golang installed. Note also that for other agent connect methods (**Launch via SSH** or **Launch via JNLP**), you will need different agent Docker imagers (jenkins/ssh-agent or jenkins/inbound-agent). For details, please check https://plugins.jenkins.io/docker-plugin/.

After saving, everything will be set up. We can run the pipeline to observe that the execution really takes place on the Docker agent, but first, let's dig a little deeper in order to understand how the Docker agents work.

Understanding dynamically provisioned Docker agents

Dynamically provisioned Docker agents can be treated as a layer over the standard agent mechanism. It changes neither the communication protocol nor how the agent is created. So, *what does Jenkins do with the Docker agent configuration we provided?*

The following diagram presents the Docker master-agent architecture we've configured:

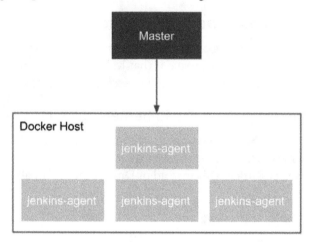

Figure 3.15 – Dynamically provisioned Docker agents

Let's describe how the Docker agent mechanism is used, step by step:

1. When the Jenkins job is started, the master runs a new container from the `jenkins/agent` image on the agent Docker host.

2. The `jenkins/agent` container starts the Jenkins agent and attaches it to the Jenkins master's nodes pool.

3. Jenkins executes the pipeline inside the `jenkins/agent` container.

4. After the build, the master stops and removes the agent container.

> **Information**
> Running the Jenkins master as a Docker container is independent of running Jenkins agents as Docker containers. It's reasonable to do both, but any of them will work separately.

The solution is somehow similar to the permanent Docker agent solution because, as a result, we run the build inside a Docker container. The difference, however, is in the agent node configuration. Here, the whole agent is dockerized – not only the build environment.

> **Tip**
> The Jenkins build usually needs to download a lot of project dependencies (for example, Gradle/Maven dependencies), which may take a lot of time. If Docker agents are automatically provisioned for each build, then it may be worthwhile to set up a Docker volume for them to enable caching between the builds.

Dynamically provisioned Kubernetes agents

We can dynamically provision agents on Kubernetes similar to how we did with the Docker host. The benefit of such an approach is that Kubernetes is a cluster of multiple physical machines that can easily scale up or down, according to needs.

Configuring dynamically provisioned Kubernetes agents

Firstly, we need to install the **Kubernetes** plugin. Then, we can follow the same steps when we installed the Docker agents. The difference starts when we click on **Add a new cloud**. This time, we need to select **Kubernetes** instead of **Docker** and fill in all the details about the Kubernetes cluster:

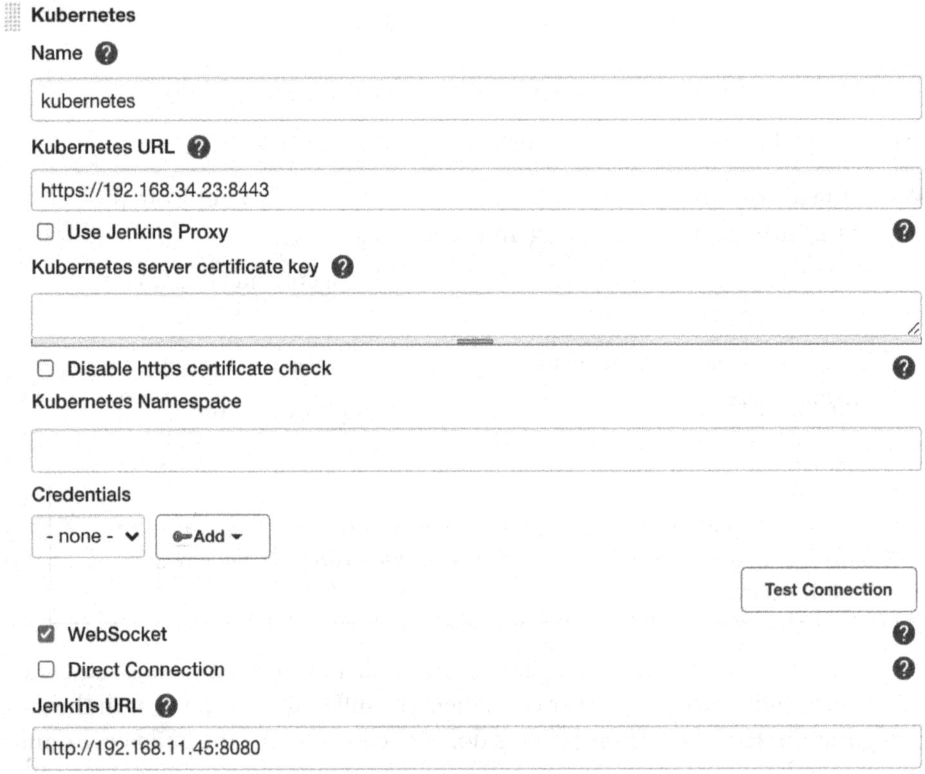

Figure 3.16 – Kubernetes agent configuration

You need to fill in **Kubernetes URL**, which is the address of your Kubernetes cluster. Usually, you will also need to enter the credentials of your Kubernetes cluster. Then, you must click on **Add Pod Template** and fill in **Pod Template** analogously to what you did for **Docker Template** in the case of the **Docker** plugin.

> **Information**
>
> For more detailed instructions on how to set up the Jenkins Kubernetes plugin, visit `https://plugins.jenkins.io/kubernetes/`.

After successful configuration, when you start a new build, Jenkins automatically provisions a new agent in Kubernetes and uses it for the pipeline execution.

> **Tip**
>
> If you install Jenkins in Kubernetes using Helm, as described at the beginning of this chapter, it is automatically configured with the Kubernetes plugin and automatically provisions Jenkins agents in the same Kubernetes cluster where the Jenkins master is deployed. This way, with one Helm command, we install a fully functional and scalable Jenkins ecosystem!

Understanding dynamically provisioned Kubernetes agents

Dynamically provisioning an agent in Kubernetes works very similarly to provisioning an agent in the Docker host. The difference is that now we interact with a cluster of machines, not just a single Docker host. This approach is presented in the following diagram:

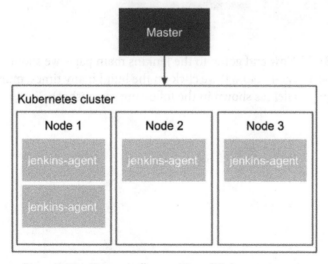

Figure 3.17 – Dynamically provisioned Kubernetes agents

Kubernetes nodes can be dynamically added and removed, which makes the whole master-agent architecture very flexible in terms of needed resources. When we experience too many Jenkins builds, we can easily add a new machine to the Kubernetes cluster and, therefore, improve the Jenkins capacity.

We have covered a lot of different strategies on how to configure Jenkins agents. Let's move on and test our configuration.

Testing agents

No matter which agent configuration you have chosen, you can now check whether everything works correctly.

Let's go back to the Hello World pipeline. Usually, the builds last longer than the Hello World example, so we can simulate it by adding `sleeping` to the pipeline script:

```
pipeline {
    agent any
    stages {
        stage("Hello") {
            steps {
                sleep 300 // 5 minutes
                echo 'Hello World'
            }
        }
    }
}
```

After clicking on **Build Now** and going to the Jenkins main page, we should see that the build is executed on an agent. Now, if we click on the build many times, multiple builds should be started in parallel (as shown in the following screenshot):

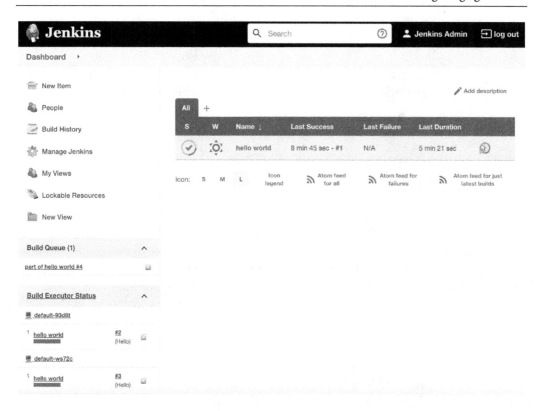

Figure 3.18 – Jenkins running multiple builds on Jenkins agents

> **Tip**
>
> To prevent job executions on the master, remember to set **# of executors** to 0
> for the master node in the **Manage Nodes** configuration.

Having seen that the agents are executing our builds, we have confirmed that they
are configured correctly. Before we move on and see how to create our own Jenkins
images, let's clarify one nuance, the difference between Docker agents and the Docker
pipeline build.

Comparing Docker pipeline builds and Docker agents

The Jenkins pipeline build executes inside a Docker container in two cases – permanent
Docker host agents and dynamically provisioned Docker/Kubernetes agents. However,
there is a subtle difference between both solutions, which requires a few words of
clarification.

Docker pipeline builds

If your agent is a Docker host, then you can specify your pipeline runtime from the Jenkins user perspective. In other words, if your project has some special build runtime requirements, you can dockerize them and describe your pipeline script as follows:

```
agent {
    docker {
        image 'custom-docker-image'
    }
}
```

Such an approach means that from the user's perspective, you are free to choose the Docker image used for your build. What's more, you can even decide to execute the build directly on the host, not inside the Docker container, which may be especially useful when the steps in your pipeline need a Docker host that may not be accessible from inside the container. We will see an example of such a requirement in the later chapters of this book.

Docker agents

If your agent itself is a Docker container, then you specify the Docker image used from the Jenkins admin perspective. In such a situation, if your project has some specific build runtime requirements, then you need to do the following:

1. Create a custom Docker image that uses `jenkins/agent` as the base image.
2. Ask a Jenkins admin to include it in the Docker/Kubernetes plugin configuration and assign a special label to the given agent.
3. Use the specific agent label inside your pipeline script.

This means that for a project with custom requirements, the setup is slightly more complex.

There is one more open question: what about a scenario when your pipeline requires access to the Docker host – for example, to build Docker images? Is there a way to use Docker inside a Docker container? Docker-in-Docker comes to the rescue.

Docker-in-Docker

There is a solution called **Docker-in-Docker** (**DIND**), which allows you to use Docker inside a Docker container. Technically, it requires granting **privileged** permissions to the Docker container, and there is a related configuration field inside the Jenkins Docker plugin. Note, however, that allowing a container to access the Docker host is a potential security hole, so you should always take extra precautions before applying such a configuration.

We have finally covered everything about the Jenkins agent configuration. Now, let's move on and look at how, and for what reasons, we can create our own Jenkins images.

Custom Jenkins images

So far, we have used Jenkins images pulled from the internet. We used `jenkins/jenkins` for the master container and `jenkins/agent` (or `jenkins/inbound-agent` or `jenkins/ssh-agent`) for the agent container. However, you may want to build your own images to satisfy the specific build environment requirements. In this section, we will cover how to do it.

Building the Jenkins agent

Let's start with the agent image because it's more frequently customized. The build execution is performed on the agent, so it's the agent that needs to have the environment adjusted to the project we want to build – for example, it may require the Python interpreter if our project is written in Python. The same applies to any library, tool, or testing framework, or anything that is needed by the project.

There are four steps to building and using the custom image:

1. Create a Docker file.
2. Build the image.
3. Push the image into a registry.
4. Change the agent configuration on the master.

As an example, let's create an agent that serves the Python project. We can build it on top of the `jenkins/agent` image, for the sake of simplicity. Let's do it using the following four steps:

1. **Dockerfile**: In a new directory, create a file named `Dockerfile`, with the following content:

    ```
    FROM jenkins/agent
    USER root
    RUN apt-get update && \
        apt-get install -y python
    USER jenkins
    ```

2. **Build the image**: We can build the image by executing the following command:

    ```
    $ docker build -t leszko/jenkins-agent-python .
    ```

3. **Push the image into a registry**: To push the image, execute the following command (if you build the image on Docker Engine that is used by the Jenkins master, you can skip this step):

    ```
    $ docker push leszko/jenkins-agent-python
    ```

    > **Tip**
    >
    > This step assumes that you have an account on Docker Hub (change `leszko` to your Docker Hub name) and that you have already executed `docker login`. We'll cover more on Docker registries in *Chapter 5, Automated Acceptance Testing*.

4. **Change the agent configuration on the master**: The last step, of course, is to set `leszko/jenkins-agent-python` instead of `jenkins/agent` in the Jenkins master's configuration (as described in the *Dynamically provisioned Docker agents* section).

    > **Tip**
    >
    > If you have pushed your image to the Docker Hub registry and the registry is private, then you'll also need to configure the appropriate credentials in the Jenkins master configuration.

What if we need Jenkins to build two different kinds of projects – for example, one based on Python and another based on Ruby? In that case, we can prepare an agent that's generic enough to support both – Python and Ruby. However, in the case of Docker, it's recommended to create a second agent image (`leszko/jenkins-agent-ruby` by analogy). Then, in the Jenkins configuration, we need to create two Docker templates and label them accordingly.

> **Information**
>
> We used `jenkins/agent` as the base image, but we can use `jenkins/inbound-agent` and `jenkins/ssh-agent` in exactly the same manner.

Building the Jenkins master

We already have a custom agent image. *Why would we also want to build our own master image?* One of the reasons might be that we don't want to use agents at all, and since the execution will be done on the master, its environment has to be adjusted to the project's needs. That is, however, a very rare case. More often, we will want to configure the master itself.

Imagine the following scenario: your organization scales Jenkins horizontally, and each team has its own instance. There is, however, some common configuration – for example, a set of base plugins, backup strategies, or the company logo. Then, repeating the same configuration for each of the teams is a waste of time. So, we can prepare the shared master image and let the teams use it.

Jenkins is natively configured using XML files, and it provides the Groovy-based DSL language to manipulate them. That is why we can add the Groovy script to the Dockerfile in order to manipulate the Jenkins configuration. Furthermore, there are special scripts to help with the Jenkins configuration if it requires something more than XML changes – for instance, plugin installation.

> **Information**
>
> All possibilities of the Dockerfile instructions are well described on the GitHub page at `https://github.com/jenkinsci/docker`.

As an example, let's create a master image with `docker-plugin` already installed and a number of executors set to 5. In order to do it, we need to perform the following:

1. Create the Groovy script to manipulate `config.xml`, and set the number of executors to 5.

2. Create the Dockerfile to install `docker-plugin`, and execute the Groovy script.

3. Build the image.

Let's use the three steps mentioned and build the Jenkins master image:

1. **Groovy script**: Create a new directory and the `executors.groovy` file with the following content:

```
import jenkins.model.*
Jenkins.instance.setNumExecutors(5)
```

> **Tip**
> The complete Jenkins API can be found on the official page at `http://javadoc.jenkins.io/`.

2. **Dockerfile**: In the same directory, create a Docker file:

```
FROM jenkins/jenkins:lts-jdk11
COPY executors.groovy /usr/share/jenkins/ref/init.groovy.d/executors.groovy
RUN jenkins-plugin-cli --plugins docker-plugin github-branch-source:1.8
```

3. **Build the image**: We can finally build the image:

```
$ docker build -t jenkins-master .
```

After the image is created, each team in the organization can use it to launch their own Jenkins instance.

> **Tip**
> Similar to the Jenkins agent image, you can build the master image as `leszko/jenkins-master` and push it into your Docker Hub account.

Having our own master and agent images lets us provide the configuration and build environment for the teams in our organization. In the next section, you'll see what else is worth being configured in Jenkins.

> **Information**
>
> You can also configure Jenkins master as well as Jenkins pipelines using the YAML-based configuration with the Configuration as Code plugin. Read more at `https://www.jenkins.io/projects/jcasc/`.

Configuration and management

We have already covered the most crucial part of the Jenkins configuration – **agent provisioning**. Since Jenkins is highly configurable, you can expect many more possibilities to adjust it to your needs. The good news is that the configuration is intuitive and accessible via the web interface, so it does not require a detailed description. Everything can be changed under the **Manage Jenkins** sub-page. In this section, we will focus on only a few aspects that are most likely to be changed – plugins, security, and backup.

Plugins

Jenkins is highly plugin-oriented, which means that a lot of features are delivered by the use of plugins. They can extend Jenkins in an almost unlimited way, which, taking into consideration the large community, is one of the reasons why Jenkins is such a successful tool. With Jenkins' openness comes risk, and it's better to download only plugins from a reliable source or check their source code.

There are literally tons of plugins to choose from. Some of them were already installed automatically, during the initial configuration. Others (Docker and Kubernetes plugins) were installed when setting the Docker agents. There are plugins for cloud integration, source control tools, code coverage, and much more. You can also write your own plugin, but it's better to check whether the one you need is already available.

> **Information**
>
> There is an official Jenkins page to browse plugins at `https://plugins.jenkins.io/`.

Security

The way you should approach Jenkins security depends on the Jenkins architecture you have chosen within your organization. If you have a Jenkins master for every small team, then you may not need it at all (under the assumption that the corporate network is firewalled). However, if you have a single Jenkins master instance for the whole organization, then you'd better be sure you've secured it well.

Jenkins comes with its own user database; we already created a user during the initial configuration process. You can create, delete, and modify users by opening the **Manage Users** setting page. The built-in database can be a solution in the case of small organizations; however, for a large group of users, you will probably want to use the **Lightweight Directory Access Protocol (LDAP)** instead. You can choose it on the **Configure Global Security** page. There, you can also assign roles, groups, and users. By default, the **Logged-in users can do anything** option is set, but in a large-scale organization, you should probably consider using more detailed permission granularity.

Backup

As the old saying goes, *there are two types of people: those who back up, and those who will back up*. Believe it or not, the backup is something you probably want to configure. *What files should be backed up, and from which machines?* Luckily, agents automatically send all the relevant data back to the master, so we don't need to bother with them. If you run Jenkins in a container, then the container itself is also not of interest, since it does not hold a persistent state. The only place we are interested in is the Jenkins home directory.

We can either install a Jenkins plugin (which will help us to set periodic backups) or simply set a cron job to archive the directory in a safe place. To reduce the size, we can exclude the subfolders that are not of interest (that will depend on your needs; however, almost certainly, you don't need to copy the following: *war*, *cache*, *tools*, and *workspace*).

> **Information**
>
> If you automate your Jenkins master setup (by building a custom Docker image or using the Jenkins Configuration as Code plugin), then you may consider skipping the Jenkins backup configuration.

Jenkins Blue Ocean UI

The first version of Hudson (the former Jenkins) was released in 2005. It's been on the market for more than 15 years now. However, its look and feel haven't changed much. We've used it for quite a while, and it's hard to deny that it looks outdated. Blue Ocean is the plugin that has redefined the user experience of Jenkins. If Jenkins is aesthetically displeasing to you or its workflow does not feel intuitive enough, then it's definitely worth giving Blue Ocean a try (as shown in the following screenshot):

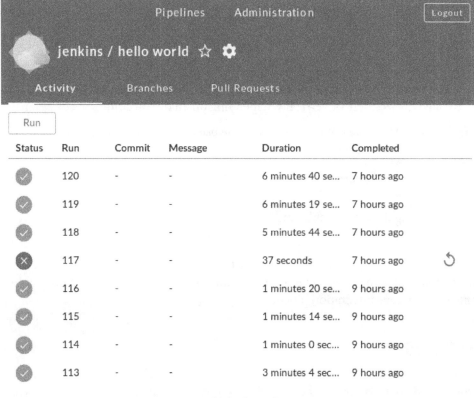

Figure 3.19 – The Jenkins Blue Ocean UI

> **Information**
>
> You can read more on the Blue Ocean page at `https://www.jenkins.io/doc/book/blueocean/`.

Summary

In this chapter, we covered the Jenkins environment and its configuration. The knowledge we have gained is sufficient to set up the complete Docker-based Jenkins infrastructure. The key takeaway points from the chapter are as follows:

- Jenkins is a general-purpose automation tool that can be used with any language or framework.

- Jenkins is highly extensible by plugins, which can be written or found on the internet.

- Jenkins is written in Java, so it can be installed on any operating system. It's also officially delivered as a Docker image.

- Jenkins can be scaled using the master-agent architecture. The master instances can be scaled horizontally or vertically, depending on an organization's needs.

- Jenkins agents can be implemented with the use of Docker, which helps in automatic configuration and dynamic agent allocation.

- Custom Docker images can be created for both the Jenkins master and Jenkins agent.

- Jenkins is highly configurable, and some aspects that should always be considered are security and backups.

In the next chapter, we will focus on something that we already touched on with the Hello World example – pipelines. We will describe the idea behind and the method for building a complete continuous integration pipeline.

Exercises

You learned a lot about Jenkins configuration throughout this chapter. To consolidate your knowledge, we recommend the following exercises on preparing Jenkins images and testing the Jenkins environment:

1. Create Jenkins master and agent Docker images and use them to run a Jenkins infrastructure capable of building Ruby projects:

    I.   Create the Jenkins master Dockerfile, which automatically installs the Docker plugin.

    II. Build the master image and run the Jenkins instance.

    III. Create the agent Dockerfile (suitable for the dynamic agent provisioning), which installs the Ruby interpreter.

IV.   Build the agent image.

V.   Change the configuration in the Jenkins instance to use the agent image.

2. Create a pipeline that runs a Ruby script printing `Hello World from Ruby`:

   I.   Create a new pipeline.

   II.   Use the following shell command to create the `hello.rb` script on the fly:

   ```
   sh "echo \"puts 'Hello World from Ruby'\" > hello.rb"
   ```

3. Add the command to run `hello.rb`, using the Ruby interpreter.

4. Run the build and observe the console's output.

Questions

To verify your knowledge from this chapter, please answer the following questions:

1. Is Jenkins provided in the form of a Docker image?

2. What is the difference between a Jenkins master and a Jenkins agent (slave)?

3. What is the difference between vertical and horizontal scaling?

4. What are the two main options for master-agent communication when starting a Jenkins agent?

5. What is the difference between setting up a permanent agent and a permanent Docker agent?

6. When would you need to build a custom Docker image for a Jenkins agent?

7. When would you need to build a custom Docker image for a Jenkins master?

8. What is Jenkins Blue Ocean?

Further reading

To read more about Jenkins, please refer to the following resources:

- *Jenkins Handbook*: https://www.jenkins.io/doc/book/

- *Jenkins Essentials, Mitesh Soni*: https://www.packtpub.com/virtualization-and-cloud/jenkins-essentials-second-edition

- *Jenkins: The Definitive Guide, John Ferguson Smart*: https://www.oreilly.com/library/view/jenkins-the-definitive/9781449311155/

Section 2 – Architecting and Testing an Application

In this section, we will cover continuous integration pipeline steps and Docker Hub registry concepts. Kubernetes will also be introduced, and we will learn how to scale a pool of Docker servers.

The following chapters are covered in this section:

- *Chapter 4, Continuous Integration Pipeline*
- *Chapter 5, Automated Acceptance Testing*
- *Chapter 6, Clustering with Kubernetes*

4

Continuous Integration Pipeline

We already know how to configure Jenkins. In this chapter, we will see how to use it effectively, focusing on the feature that lies at the heart of Jenkins – pipelines. By building a complete continuous integration process from scratch, we will describe all aspects of modern team-oriented code development.

This chapter covers the following topics:

- Introducing pipelines
- The commit pipeline
- Code quality stages
- Triggers and notifications
- Team development strategies

Technical requirements

To complete this chapter, you'll need the following software:

- Jenkins
- Java JDK 8+

All the examples and solutions to the exercises can be found at `https://github.com/PacktPublishing/Continuous-Delivery-With-Docker-and-Jenkins-3rd-Edition/tree/main/Chapter04`.

Code in Action videos for this chapter can be viewed at `https://bit.ly/3r9lbmG`.

Introducing pipelines

A **pipeline** is a sequence of automated operations that usually represents a part of the software delivery and quality assurance process. It can be seen as a chain of scripts that provide the following additional benefits:

- **Operation grouping**: Operations are grouped together into stages (also known as **gates** or **quality gates**) that introduce a structure into a process and clearly define a rule – if one stage fails, no further stages are executed.
- **Visibility**: All aspects of a process are visualized, which helps in quick failure analysis and promotes team collaboration.
- **Feedback**: Team members learn about problems as soon as they occur so that they can react quickly.

> **Information**
> The concept of pipelining is similar to most continuous integration tools. However, the naming can differ. In this book, we will stick to the Jenkins terminology.

Let's first describe the Jenkins pipeline structure and then how it works in action.

The pipeline structure

A Jenkins pipeline consists of two kinds of elements – a **stage** and a **step**. The following diagram shows how they are used:

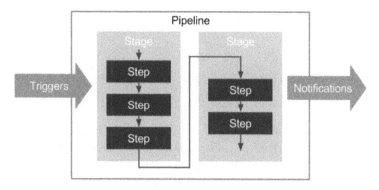

Figure 4.1 – The Jenkins pipeline structure

The following are the basic pipeline elements:

- **Step**: A single operation that tells Jenkins what to do – for example, check out code from the repository and execute a script

- **Stage**: A logical separation of steps that groups conceptually distinct sequences of steps – for example, **build**, **test**, and **deploy**, used to visualize the Jenkins pipeline progress

> **Information**
> Technically, it's possible to create parallel steps; however, it's better to treat them as an exception that is only used for optimization purposes.

A multi-stage Hello World

As an example, let's extend the `Hello World` pipeline to contain two stages:

```
pipeline {
    agent any
    stages {
        stage('First Stage') {
            steps {
                echo 'Step 1. Hello World'
            }
        }
        stage('Second Stage') {
            steps {
                echo 'Step 2. Second time Hello'
```

```
                        echo 'Step 3. Third time Hello'
                }
        }
    }
}
```

The pipeline has no special requirements in terms of environment, and it executes three steps inside two stages. When we click on **Build Now**, we should see a visual representation:

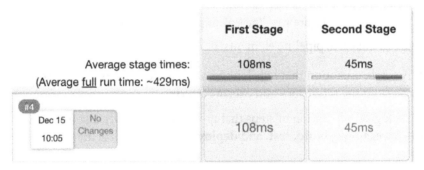

Figure 4.2 – The multi-stage pipeline build

The pipeline succeeded, and we can see the step execution details by clicking on the console. If any of the steps failed, processing would stop, and no further steps would run. Actually, the sole reason for a pipeline is to prevent all further steps from execution and visualize the point of failure.

The pipeline syntax

We've discussed the pipeline elements and already used a few of the pipeline steps – for example, echo. *What other operations can we use inside the pipeline definition?*

> **Information**
>
> In this book, we use the declarative syntax that is recommended for all new projects. The other options are a Groovy-based DSL and (prior to Jenkins 2) XML (created through the web interface).

The declarative syntax was designed to make it as simple as possible to understand the pipeline, even by people who do not write code on a daily basis. This is why the syntax is limited only to the most important keywords.

Let's try an experiment, but before we describe all the details, please read the following pipeline definition and try to guess what it does:

```
pipeline {
        agent any
        triggers { cron('* * * * *') }
        options { timeout(time: 5) }
        parameters {
                booleanParam(name: 'DEBUG_BUILD', defaultValue: true,
                description: 'Is it the debug build?')
        }
        stages {
                stage('Example') {
                        environment { NAME = 'Rafal' }
                        when { expression { return params.DEBUG_BUILD }
}
                        steps {
                                echo "Hello from $NAME"
                                script {
                                        def browsers = ['chrome', 'firefox']
                                        for (int i = 0; i < browsers.size();
++i) {
                                                echo "Testing the ${browsers[i]}
browser."
                                        }
                                }
                        }
                }
        }
        post { always { echo 'I will always say Hello again!' } }
}
```

Hopefully, the pipeline didn't scare you. It is quite complex. Actually, it is so complex that it contains most available Jenkins instructions. To answer the experiment puzzle, let's see what the pipeline does instruction by instruction:

1. Uses any available agent
2. Executes automatically every minute

3. Stops if the execution takes more than 5 minutes

4. Asks for the Boolean input parameter before starting

5. Sets `Rafal` as the `NAME` environment variable

6. Does the following, only in the case of the `true` input parameter:

    - Prints `Hello from Rafal`

    - Prints `Testing the chrome browser`

    - Prints `Testing the firefox browser`

7. Prints `I will always say Hello again!`, regardless of whether there are any errors during the execution

Now, let's describe the most important Jenkins keywords. A declarative pipeline is always specified inside the `pipeline` block and contains sections, directives, and steps. We will walk through each of them.

Information

The complete pipeline syntax description can be found on the official Jenkins page at `https://jenkins.io/doc/book/pipeline/syntax/`.

Sections

Sections define the pipeline structure and usually contain one or more directives or steps. They are defined with the following keywords:

- **Stages**: This defines a series of one or more stage directives.

- **Steps**: This defines a series of one or more step instructions.

- **Post**: This defines a series of one or more step instructions that are run at the end of the pipeline build; they are marked with a condition (for example, always, success, or failure) and are usually used to send notifications after the pipeline build (we will cover this in detail in the *Triggers and notifications* section).

- **Agent**: This specifies where the execution takes place and can define `label` to match the equally labeled agents, or `docker` to specify a container that is dynamically provisioned to provide an environment for the pipeline execution.

Directives

Directives express the configuration of a pipeline or its parts:

- **Triggers**: This defines automated ways to trigger the pipeline and can use `cron` to set the time-based scheduling, or `pollSCM` to check the repository for changes (we will cover this in detail in the *Triggers and notifications* section).

- **Options**: This specifies pipeline-specific options – for example, `timeout` (the maximum time of a pipeline run) or `retry` (the number of times the pipeline should be rerun after failure).

- **Environment**: This defines a set of key values used as environment variables during the build.

- **Parameters**: This defines a list of user-input parameters.

- **Stage**: This allows for the logical grouping of steps.

- **When**: This determines whether the stage should be executed, depending on the given condition.

- **Tools**: This defines the tools to install and put on `PATH`.

- **Input**: This allows us to prompt the input parameters.

- **Parallel**: This allows us to specify stages that are run in parallel.

- **Matrix**: This allows us to specify combinations of parameters for which the given stages run in parallel.

Steps

Steps are the most fundamental part of the pipeline. They define the operations that are executed, so they actually tell Jenkins *what to do*:

- `sh`: This executes the shell command; actually, it's possible to define almost any operation using `sh`.

- `custom`: Jenkins offers a lot of operations that can be used as steps (for example, `echo`); many of them are simply wrappers over the `sh` command used for convenience. Plugins can also define their own operations.

- `script`: This executes a block of Groovy-based code that can be used for some non-trivial scenarios where flow control is needed.

> **Information**
>
> The complete specification of the available steps can be found at `https://jenkins.io/doc/pipeline/steps/`.

Note that the pipeline syntax is very generic and, technically, can be used for almost any automation process. This is why the pipeline should be treated as a method of structuring and visualization. However, the most common use case is to implement the continuous integration server, which we will look at in the following section.

The commit pipeline

The most basic continuous integration process is called a **commit pipeline**. This classic phase, as its name indicates, starts with `commit` (or `push` in Git) to the main repository and results in a report about the build success or failure. Since it runs after each change in the code, the build should take no more than 5 minutes and should consume a reasonable amount of resources. The commit phase is always the starting point of the continuous delivery process and provides the most important feedback cycle in the development process – constant information if the code is in a healthy state.

The commit phase works as follows: a developer checks in the code to the repository, the continuous integration server detects the change, and the build starts. The most fundamental commit pipeline contains three stages:

- **Checkout**: This stage downloads the source code from the repository.
- **Compile**: This stage compiles the source code.
- **Unit test**: This stage runs a suite of unit tests.

Let's create a sample project and see how to implement the commit pipeline.

> **Information**
>
> This is an example of a pipeline for a project that uses technologies such as Git, Java, Gradle, and Spring Boot. Nevertheless, the same principles apply to any other technology.

Checkout

Checking out code from the repository is always the first operation in any pipeline. In order to see this, we need to have a repository. Then, we are able to create a pipeline.

Creating a GitHub repository

Creating a repository on the GitHub server takes just a few steps:

1. Go to `https://github.com/`.
2. Create an account if you don't have one yet.
3. Click on **New**, next to **Repositories**.
4. Give it a name – `calculator`.
5. Tick **Initialize this repository with a README**.
6. Click on **Create repository**.

Now, you should see the address of the repository – for example, `https://github.com/leszko/calculator.git`.

Creating a checkout stage

We can create a new pipeline called `calculator`, and as it is a **pipeline script**, place the code with a stage called `Checkout`:

```
pipeline {
    agent any
    stages {
        stage("Checkout") {
            steps {
                git url: 'https://github.com/leszko/calculator.git', branch: 'main'
            }
        }
    }
}
```

The pipeline can be executed on any of the agents, and its only step does nothing more than download code from the repository. We can click on **Build Now** to see whether it was executed successfully.

> **Information**
> The Git toolkit needs to be installed on the node where the build is executed.

When we have the checkout, we're ready for the second stage.

Compile

In order to compile a project, we need to do the following:

1. Create a project with the source code.

2. Push it to the repository.

3. Add the `Compile` stage to the pipeline.

Let's look at these steps in detail.

Creating a Java Spring Boot project

Let's create a very simple Java project using the Spring Boot framework built by Gradle.

> **Information**
>
> Spring Boot is a Java framework that simplifies building enterprise applications. Gradle is a build automation system that is based on the concepts of Apache Maven.

The simplest way to create a Spring Boot project is to perform the following steps:

1. Go to `http://start.spring.io/`.

2. Select **Gradle Project** instead of **Maven Project** (you can choose Maven if you prefer it to Gradle).

3. Fill **Group** and **Artifact** (for example, `com.leszko` and `calculator`).

4. Add **Web** to **Dependencies**.

5. Click on **Generate**.

6. The generated skeleton project should be downloaded (the `calculator.zip` file).

The following screenshot shows the `http://start.spring.io/` page:

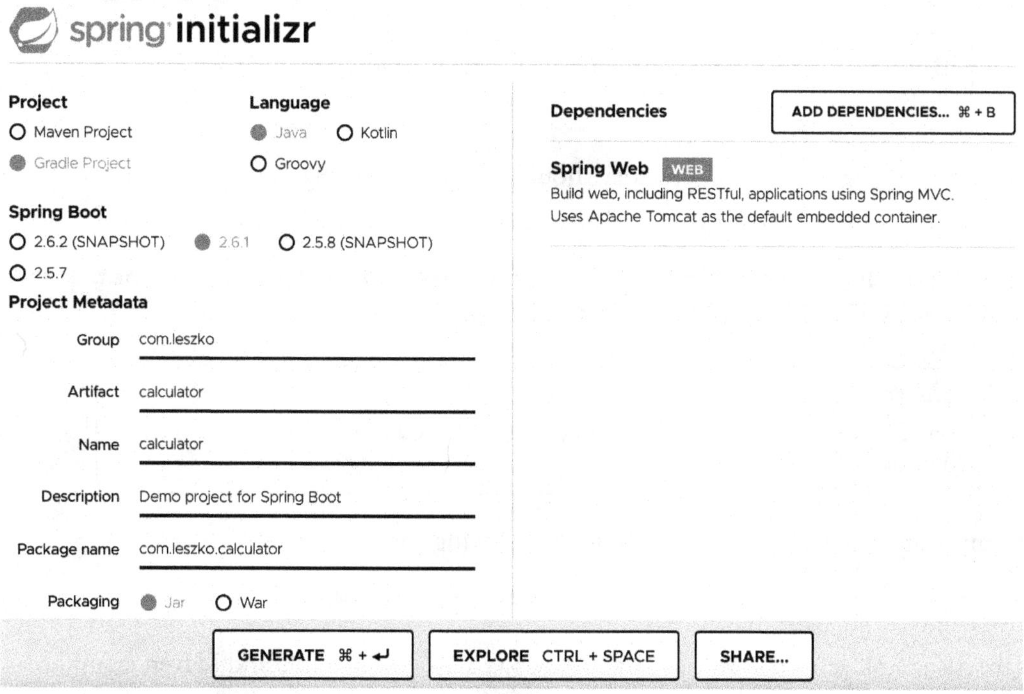

Figure 4.3 – spring initializr

After the project is created, we can push it into the GitHub repository.

Pushing code to GitHub

We will use the Git tool to perform the commit and push operations.

> **Information**
>
> In order to run the git command, you need to have the Git toolkit installed (it can be downloaded from https://git-scm.com/downloads).

Let's first clone the repository to the filesystem:

```
$ git clone https://github.com/leszko/calculator.git
```

Extract the project downloaded from http://start.spring.io/ into the directory created by Git.

> **Tip**
>
> If you prefer, you can import the project into IntelliJ, Visual Studio Code, Eclipse, or your favorite IDE tool.

As a result, the `calculator` directory should have the following files:

```
$ ls -a
. .. build.gradle .git .gitignore gradle gradlew gradlew.bat
HELP.md README.md settings.gradle src
```

> **Information**
>
> In order to perform the Gradle operations locally, you need to have the Java JDK installed.

We can compile the project locally using the following code:

```
$ ./gradlew compileJava
```

In the case of Maven, you can run `./mvnw compile`. Both Gradle and Maven compile the Java classes located in the `src` directory.

Now, we can commit and push to the GitHub repository:

```
$ git add .
$ git commit -m "Add Spring Boot skeleton"
$ git push -u origin main
```

The code is already in the GitHub repository. If you want to check it, you can go to the GitHub page and see the files.

Creating a Compile stage

We can add a `Compile` stage to the pipeline using the following code:

```
stage("Compile") {
    steps {
        sh "./gradlew compileJava"
    }
}
```

Note that we used exactly the same command locally and in the Jenkins pipeline, which is a very good sign because the local development process is consistent with the continuous integration environment. After running the build, you should see two green boxes. You can also check that the project was compiled correctly in the console log.

Unit tests

It's time to add the last stage, which is the unit test; it checks whether our code does what we expect it to do. We have to do the following:

1. Add the source code for the calculator logic.

2. Write a unit test for the code.

3. Add a Jenkins stage to execute the unit test.

Let's elaborate more on these steps next.

Creating business logic

The first version of the calculator will be able to add two numbers. Let's add the business logic as a class in the `src/main/java/com/leszko/calculator/Calculator.java` file:

```
package com.leszko.calculator;
import org.springframework.stereotype.Service;


@Service
public class Calculator {
    public int sum(int a, int b) {
        return a + b;
    }
}
```

To execute the business logic, we also need to add the web service controller in a separate file: `src/main/java/com/leszko/calculator/CalculatorController.java`:

```
package com.leszko.calculator;
import org.springframework.beans.factory.annotation.Autowired;
import org.springframework.web.bind.annotation.RequestMapping;
import org.springframework.web.bind.annotation.RequestParam;
```

```
import org.springframework.web.bind.annotation.RestController;

@RestController
class CalculatorController {
    @Autowired
    private Calculator calculator;

    @RequestMapping("/sum")
    String sum(@RequestParam("a") Integer a,
            @RequestParam("b") Integer b) {
        return String.valueOf(calculator.sum(a, b));
    }
}
```

This class exposes business logic as a web service. We can run the application and see how it works:

```
$ ./gradlew bootRun
```

This should start our web service, and we can check that it works by navigating to the browser and opening http://localhost:8080/sum?a=1&b=2. This should sum two numbers (1 and 2) and show 3 in the browser.

Writing a unit test

We already have the working application. *How can we ensure that the logic works as expected?* We tried it once, but in order to know that it will work consistently, we need a unit test. In our case, it will be trivial, maybe even unnecessary; however, in real projects, unit tests can save you from bugs and system failures.

Let's create a unit test in the src/test/java/com/leszko/calculator/ CalculatorTest.java file:

```
package com.leszko.calculator;
import org.junit.Test;
import static org.junit.Assert.assertEquals;

public class CalculatorTest {
    private Calculator calculator = new Calculator();
```

```
@Test
public void testSum() {
        assertEquals(5, calculator.sum(2, 3));
    }
}
```

Our test uses the JUnit library, so we need to add it as a dependency in the `build.gradle` file:

```
dependencies {
    ...
testImplementation 'junit:junit:4.13'
}
```

We can run the test locally using the `./gradlew test` command. Then, let's commit the code and push it to the repository:

```
$ git add .
$ git commit -m "Add sum logic, controller and unit test"
$ git push
```

Creating a Unit test stage

Now, we can add a `Unit test` stage to the pipeline:

```
stage("Unit test") {
    steps {
        sh "./gradlew test"
    }
}
```

> **Tip**
> In the case of Maven, use the `./mvnw test` command instead.

When we build the pipeline again, we should see three boxes, which means that we've completed the continuous integration pipeline:

	Checkout	Compile	Unit test
Average stage times: (Average full run time: ~22s)	579ms	25s	12s
#3 Dec 15 1 21:48 commit	623ms	5s	12s

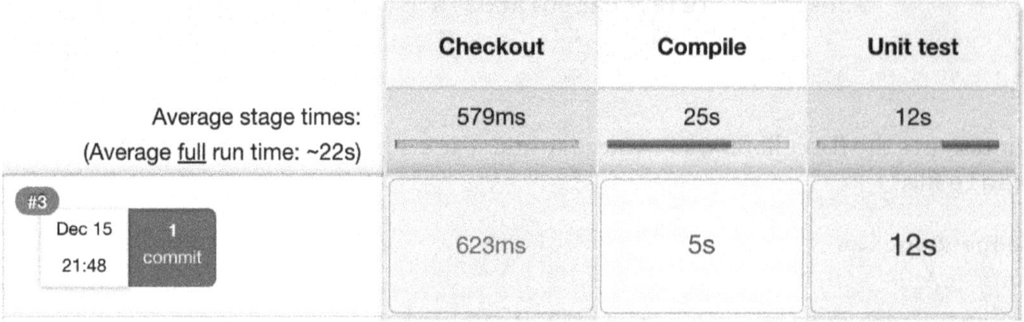

Figure 4.4 – A continuous integration pipeline build

Now that we have our pipeline prepared, let's look at how to achieve exactly the same result using Jenkinsfile.

Jenkinsfile

So far, we've created all the pipeline code directly in Jenkins. This is, however, not the only option. We can also put the pipeline definition inside a file called Jenkinsfile and commit it to the repository, together with the source code. This method is even more consistent because the way your pipeline looks is strictly related to the project itself.

For example, if you don't need the code compilation because your programming language is interpreted (and not compiled), you won't have the Compile stage. The tools you use also differ, depending on the environment. We used Gradle/Maven because we've built a Java project; however, in the case of a project written in Python, you can use PyBuilder. This leads to the idea that the pipelines should be created by the same people who write the code – the developers. Also, the pipeline definition should be put together with the code, in the repository.

This approach brings immediate benefits, as follows:

- In the case of a Jenkins failure, the pipeline definition is not lost (because it's stored in the code repository, not in Jenkins).
- The history of the pipeline changes is stored.

- Pipeline changes go through the standard code development process (for example, they are subjected to code reviews).

- Access to the pipeline changes is restricted in exactly the same way as access to the source code.

Let's see how it all looks in practice by creating a `Jenkinsfile` file.

Creating the Jenkins file

We can create the `Jenkinsfile` file and push it into our GitHub repository. Its content is almost the same as the commit pipeline we wrote. The only difference is that the checkout stage becomes redundant because Jenkins has to first check out the code (together with `Jenkinsfile`) and then read the pipeline structure (from `Jenkinsfile`). This is why Jenkins needs to know the repository address before it reads `Jenkinsfile`.

Let's create a file called `Jenkinsfile` in the `root` directory of our project:

```
pipeline {
    agent any
    stages {
        stage("Compile") {
            steps {
                sh "./gradlew compileJava"
            }
        }
        stage("Unit test") {
            steps {
                sh "./gradlew test"
            }
        }
    }
}
```

We can now commit the added files and push them to the GitHub repository:

```
$ git add Jenkinsfile
$ git commit -m "Add Jenkinsfile"
$ git push
```

Running the pipeline from Jenkinsfile

When `Jenkinsfile` is in the repository, all we have to do is to open the pipeline configuration and do the following in the `Pipeline` section:

1. Change **Definition** from **Pipeline script** to **Pipeline script from SCM**.

2. Select **Git** in **SCM**.

3. Put `https://github.com/leszko/calculator.git` in **Repository URL**.

4. Use `*/main` as **Branch Specifier**.

Figure 4.5 – The Jenkinsfile pipeline configuration

After saving, the build will always run from the current version of `Jenkinsfile` in the repository.

We have successfully created the first complete commit pipeline. It can be treated as a minimum viable product, and actually, in many cases, this suffices as the continuous integration process. In the following sections, we will see what improvements can be done to make the commit pipeline even better.

Code-quality stages

We can extend the three classic steps of continuous integration with additional steps. The most popular are code coverage and static analysis. Let's look at each of them.

Code coverage

Think about the following scenario: you have a well-configured continuous integration process; however, nobody in your project writes unit tests. It passes all the builds, but it doesn't mean that the code is working as expected. *What do we do then? How do we ensure that the code is tested?*

The solution is to add a code coverage tool that runs all tests and verifies which parts of the code have been executed. Then, it can create a report that shows the untested sections. Moreover, we can make the build fail when there is too much untested code.

There are a lot of tools available to perform the test coverage analysis; for Java, the most popular are JaCoCo, OpenClover, and Cobertura.

Let's use JaCoCo and show how the coverage check works. In order to do this, we need to perform the following steps:

1. Add JaCoCo to the Gradle configuration.
2. Add the code coverage stage to the pipeline.
3. Optionally, publish JaCoCo reports in Jenkins.

Let's look at these steps in detail.

Adding JaCoCo to Gradle

In order to run JaCoCo from Gradle, we need to add the `jacoco` plugin to the `build.gradle` file by inserting the following line:

```
plugins {
    ...
    id 'jacoco'
}
```

Next, if we want to make Gradle fail in the case of low code coverage, we can add the following configuration to the `build.gradle` file:

```
jacocoTestCoverageVerification {
    violationRules {
```

```
        rule {
            limit {
                minimum = 0.2
            }
        }
    }
}
```

This configuration sets the minimum code coverage to 20%. We can run it with the following command:

```
$ ./gradlew test jacocoTestCoverageVerification
```

This command checks whether the code coverage is at least 20%. You can play with the minimum value to see the level at which the build fails. We can also generate a test coverage report using the following command:

```
$ ./gradlew test jacocoTestReport
```

You can check out the full coverage report in the `build/reports/jacoco/test/html/index.html` file:

calculator

Element	Missed Instructions	Cov.	Missed Branches	Cov.	Missed	Cxty	Missed	Lines
com.leszko.calculator		33%		n/a	3	6	4	7
Total	18 of 27	33%	0 of 0	n/a	3	6	4	7

Created with JaCoCo 0.8.7.202105040129

Figure 4.6 – JaCoCo code coverage report

Let's now add the coverage stage in our pipeline.

Adding a code coverage stage

Adding a code coverage stage to the pipeline is as simple as the previous stages:

```
stage("Code coverage") {
    steps {
        sh "./gradlew jacocoTestReport"
        sh "./gradlew jacocoTestCoverageVerification"
    }
}
```

After adding this stage, if anyone commits code that is not well covered with tests, the build will fail.

Publishing the code coverage report

When coverage is low and the pipeline fails, it is useful to look at the code coverage report and find what parts are not yet covered with tests. We can run Gradle locally and generate the coverage report; however, it is more convenient if Jenkins shows the report for us.

In order to publish the code coverage report in Jenkins, we require the following stage definition:

```
stage("Code coverage") {
    steps {
        sh "./gradlew jacocoTestReport"
        publishHTML (target: [
            reportDir: 'build/reports/jacoco/test/html',
            reportFiles: 'index.html',
            reportName: "JaCoCo Report"
        ])
        sh "./gradlew jacocoTestCoverageVerification"
    }
}
```

This stage copies the generated JaCoCo report to the Jenkins output. When we run the build again, we should see a link to the code coverage reports (in the menu on the left-hand side, below **Build Now**).

Information

To perform the publishHTML step, you need to have the HTML Publisher plugin installed in Jenkins. You can read more about the plugin at https://www.jenkins.io/doc/pipeline/steps/htmlpublisher/. Note also that if the report is generated but not displayed properly in Jenkins, you may need to configure Jenkins Security, as described here: https://www.jenkins.io/doc/book/security/configuring-content-security-policy/.

We have created the code coverage stage, which shows the code that is not tested and therefore vulnerable to bugs. Let's see what else can be done in order to improve the code quality.

> **Tip**
>
> If you need code coverage that is stricter, you can check the concept of mutation testing and add the PIT framework stage to the pipeline. Read more at `http://pitest.org/`.

Static code analysis

Your code coverage may work perfectly fine; however, *what about the quality of the code itself? How do we ensure it is maintainable and written in a good style?*

Static code analysis is an automatic process of checking code without actually executing it. In most cases, it implies checking a number of rules on the source code. These rules may apply to a wide range of aspects; for example, all public classes need to have a Javadoc comment, the maximum length of a line is 120 characters, or if a class defines the `equals()` method, it has to define the `hashCode()` method as well.

The most popular tools to perform static analysis on Java code are Checkstyle, FindBugs, and PMD. Let's look at an example and add the static code analysis stage using Checkstyle. We will do this in three steps:

1. Adding the Checkstyle configuration
2. Adding the Checkstyle stage
3. Optionally, publishing the Checkstyle report in Jenkins

We will walk through each of them.

Adding the Checkstyle configuration

In order to add the Checkstyle configuration, we need to define the rules against which the code is checked. We can do this by specifying the `config/checkstyle/checkstyle.xml` file:

```
<?xml version="1.0"?>
<!DOCTYPE module PUBLIC
        "-//Puppy Crawl//DTD Check Configuration 1.2//EN"
        "http://www.puppycrawl.com/dtds/configuration_1_2.dtd">
```

```
<module name="Checker">
    <module name="TreeWalker">
        <module name="ConstantName" />
    </module>
</module>
```

The configuration contains only one rule – checking whether all Java constants follow the naming convention and consist of uppercase characters only.

> **Information**
> The complete Checkstyle description can be found at `https://checkstyle.sourceforge.io/config.html`.

We also need to add the `checkstyle` plugin to the `build.gradle` file:

```
plugins {
    ...
    id 'checkstyle'
}
```

Then, we can run `checkstyle` with the following command:

```
$ ./gradlew checkstyleMain
```

In the case of our project, this command should complete successfully because we didn't use any constants so far. However, you can try adding a constant with the wrong name and checking whether the build fails. For example, if you add the following constant to the `src/main/java/com/leszko/calculator/CalculatorApplication.java` file, `checkstyle` fails:

```
@SpringBootApplication
public class CalculatorApplication {
    private static final String constant = "constant";
    public static void main(String[] args) {
        SpringApplication.run(CalculatorApplication.class,
args);
    }
}
```

Adding a Static code analysis stage

We can add a `Static code analysis` stage to the pipeline:

```
stage("Static code analysis") {
    steps {
        sh "./gradlew checkstyleMain"
    }
}
```

Now, if anyone commits any code that does not follow the Java constant naming convention, the build will fail.

Publishing static code analysis reports

Very similar to JaCoCo, we can add the Checkstyle report to Jenkins:

```
publishHTML (target: [
    reportDir: 'build/reports/checkstyle/',
    reportFiles: 'main.html',
    reportName: "Checkstyle Report"
])
```

This generates a link to the Checkstyle report.

We have now added the static code analysis stage, which can help to find bugs and standardize the code style inside a team or organization.

Let's see one more option we have when it comes to implementing code analysis.

SonarQube

SonarQube is the most widespread source code quality management tool.
It supports multiple programming languages and can be treated as an alternative to the code coverage and static code analysis steps we looked at. Actually, it is a separate server that aggregates different code analysis frameworks, such as Checkstyle, FindBugs, and JaCoCo. It has its own dashboards and integrates well with Jenkins.

Instead of adding code quality steps to the pipeline, we can install SonarQube, add plugins there, and add a *sonar* stage to the pipeline. The advantage of this solution is that SonarQube provides a user-friendly web interface to configure rules and show code vulnerabilities.

> **Information**
>
> You can read more about SonarQube on its official page at `https://www.sonarqube.org/`.

Now that we have covered the code quality stages, let's focus on triggers and notifications.

Triggers and notifications

So far, we have always built the pipeline manually by clicking on the **Build Now** button. It works completely fine but may not be very convenient in practice. All team members would have to remember that after committing to the repository, they need to open Jenkins and start the build. The same applies to pipeline monitoring; so far, we have manually opened Jenkins and checked the build status. In this section, we will see how to improve the process so that the pipeline will start automatically and, when completed, notify team members regarding its status.

Triggers

An automatic action to start the build is called the pipeline trigger. In Jenkins, there are many options to choose from; however, they all boil down to three types:

- External
- Polling **Source Control Management (SCM)**
- A scheduled build

Let's take a look at each of them.

External

External triggers are easy to understand. They mean that Jenkins starts the build after it's called by the **notifier**, which can be the other pipeline build, the SCM system (for example, GitHub), or any remote script.

The following diagram presents the communication:

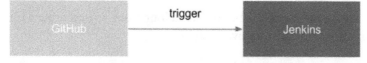

Figure 4.7 – An external trigger

GitHub triggers Jenkins after a push to the repository and the build is started.

To configure the system this way, we need the following setup steps:

1. Install the GitHub plugin in Jenkins.

2. Generate a secret key for Jenkins.

3. Set the GitHub webhook and specify the Jenkins address and key.

In the case of the most popular SCM providers, dedicated Jenkins plugins are always provided.

There is also a more generic way to trigger Jenkins via the REST call to the `<jenkins_url>/job/<job_name>/build?token=<token>` endpoint. For security reasons, it requires setting `token` in Jenkins and then using it in the remote script.

> **Information**
>
> Jenkins must be accessible from the SCM server. In other words, if we use the public GitHub repository to trigger Jenkins, our Jenkins server must be public as well. This also applies to the REST call solution, in which case, the `<jenkins_url>` address must be accessible from the script that triggers it.

Polling SCM

Polling the SCM trigger is a little less intuitive. The following diagram presents the communication:

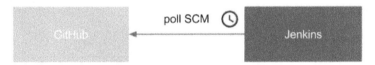

Figure 4.8 – Polling the SCM trigger

Jenkins periodically calls GitHub and checks whether there was any push to the repository. Then, it starts the build. It may sound counter-intuitive, but there are at least two good cases for using this method:

*   Jenkins is inside the firewalled network (which GitHub does not have access to).

*   Commits are frequent and the build takes a long time, so executing a build after every commit would cause an overload.

The configuration of `pollSCM` is also somehow simpler because the way to connect from Jenkins to GitHub is already set up (Jenkins checks out the code from GitHub, so it knows how to access it). In the case of our calculator project, we can set up an automatic trigger by adding the `triggers` declaration (just after `agent`) to the pipeline:

```
triggers {
    pollSCM('* * * * *')
}
```

After running the pipeline manually for the first time, the automatic trigger is set. Then, it checks GitHub every minute, and for new commits, it starts a build. To test that it works as expected, you can commit and push anything to the GitHub repository and see that the build starts.

We used the mysterious `* * * * *` as an argument to `pollSCM`. It specifies how often Jenkins should check for new source changes and is expressed in the `cron`-style string format.

> **Information**
>
> The `cron` string format is described (together with the cron tool) at https://en.wikipedia.org/wiki/Cron.

Scheduled builds

The scheduled trigger means that Jenkins runs the build periodically, regardless of whether there were any commits to the repository.

As the following screenshot shows, no communication with any system is required:

Figure 4.9 – The scheduled build trigger

The implementation of **Scheduled build** is exactly the same as polling SCM. The only difference is that the `cron` keyword is used instead of `pollSCM`. This trigger method is rarely used for the commit pipeline but applies well to nightly builds (for example, complex integration testing executed at night).

Notifications

Jenkins provides a lot of ways to announce its build status. What's more, as with everything in Jenkins, new notification types can be added using plugins.

Let's walk through the most popular types so that you can choose the one that fits your needs.

Email

The most classic way to notify users about the Jenkins build status is to send emails. The advantage of this solution is that everybody has a mailbox, everybody knows how to use it, and everybody is used to receiving information in it. The drawback is that, usually, there are simply too many emails, and the ones from Jenkins quickly become filtered out and never read.

The configuration of the email notification is very simple:

1. Have the **SMTP (Simple Mail Transfer Protocol)** server configured.
2. Set its details in Jenkins (in **Manage Jenkins | Configure System**).
3. Use the `mail to` instruction in the pipeline.

The pipeline configuration can be as follows:

```
post {
    always {
        mail to: 'team@company.com',
        subject: "Completed Pipeline: ${currentBuild.
fullDisplayName}",
        body: "Your build completed, please check: ${env.
BUILD_URL}"
    }
}
```

Note that all notifications are usually called in the `post` section of the pipeline, which is executed after all steps, no matter whether the build succeeded or failed. We used the `always` keyword; however, there are different options:

- `always`: Execute regardless of the completion status.
- `changed`: Execute only if the pipeline changed its status.
- `fixed`: Execute only if the pipeline changed its status from failed to success.

- `regression`: Execute only if the pipeline changed its status from success to failed, unstable, or aborted.

- `aborted`: Execute only if the pipeline was manually aborted.

- `failure`: Execute only if the pipeline has the `failed` status.

- `success`: Execute only if the pipeline has the `success` status.

- `unstable`: Execute only if the pipeline has the `unstable` status (usually caused by test failures or code violations).

- `unsuccessful`: Execute only if the pipeline has any status other than `success`.

Group chats

If a group chat (for example, Slack) is the first method of communication in your team, it's worth considering adding the automatic build notifications there. No matter which tool you use, the procedure to configure it is always the same:

1. Find and install the plugin for your group chat tool (for example, the **Slack Notification** plugin).

2. Configure the plugin (the server URL, channel, authorization token, and so on).

3. Add the sending instruction to the pipeline.

Let's see a sample pipeline configuration for Slack to send notifications after the build fails:

```
post {
    failure {
        slackSend channel: '#dragons-team',
        color: 'danger',
        message: "The pipeline ${currentBuild.
fullDisplayName} failed."
    }
}
```

Team spaces

Together with the agile culture came the idea that it's better to have everything happening in a team space. Instead of writing emails, meet together; instead of online messaging, come and talk; instead of task tracking tools, have a whiteboard. The same idea came to continuous delivery and Jenkins. Currently, it's very common to install big screens (also called **build radiators**) in the team space. Then, when you come to the office, the first thing you see is the current status of the pipeline. Build radiators are considered one of the most effective notification strategies. They ensure that everyone is aware of failing builds and, as a beneficial side effect, they boost team spirit and favor in-person communication.

Since developers are creative beings, they invented a lot of other ideas that play the same role as the radiators. Some teams hang large speakers that beep when the pipeline fails. Others have toys that blink when the build is done. One of my favorites is Pipeline State UFO, which is provided as an open source project on GitHub. On its page, you can find a description of how to print and configure a UFO that hangs off the ceiling and signals the pipeline state. You can find more information at `https://github.com/Dynatrace/ufo`.

> **Information**
>
> Since Jenkins is extensible by plugins, its community wrote a lot of different ways to inform users about the build statuses. Among them, you can find RSS feeds, SMS notifications, mobile applications, and desktop notifiers.

Now that we have covered triggers and notifications, let's focus on one more important aspect – team development strategies.

Team development strategies

We have covered everything regarding how the continuous integration pipeline should look. However, *when exactly should it be run?* Of course, it is triggered after the commit to the repository, but *after the commit to which branch? Only to the trunk or to every branch? Or, maybe it should run before, not after, committing so that the repository will always be healthy? Or, how about the crazy idea of having no branches at all?*

There is no single best answer to these questions. Actually, the way you use the continuous integration process depends on your team development workflow. So, before we go any further, let's describe the possible workflows.

Development workflows

A development workflow is the way your team puts code into the repository. It depends, of course, on many factors, such as the SCM tool, the project specifics, and the team size.

As a result, each team develops the code in a slightly different manner. We can, however, classify them into three types: a **trunk-based workflow**, a **branching workflow**, and a **forking workflow**.

Information

All workflows are described in detail, with examples, at `https://www.atlassian.com/git/tutorials/comparing-workflows`.

The trunk-based workflow

The trunk-based workflow is the simplest possible strategy. It is presented in the following diagram:

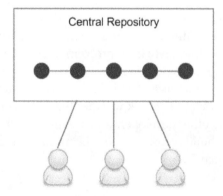

Figure 4.10 – The trunk-based workflow

There is one central repository with a single entry for all changes to the project, which is called the **trunk** or **master**. Every member of the team clones the central repository to have their own local copies. The changes are committed directly to the central repository.

The branching workflow

The branching workflow, as its name suggests, means that the code is kept in many different branches. The idea is presented in the following diagram:

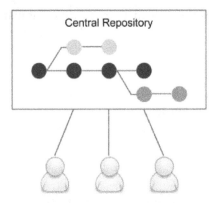

Figure 4.11 – The branching workflow

When developers start to work on a new feature, they create a dedicated branch from the trunk and commit all feature-related changes there. This makes it easy for multiple developers to work on a feature without breaking the main code base. This is why, in the case of the branching workflow, there is no problem in keeping the master healthy. When the feature is completed, a developer rebases the feature branch from the master and creates a pull request that contains all feature-related code changes. It opens the code review discussions and makes space to check whether the changes disturb the master. When the code is accepted by other developers and automatic system checks, it is merged into the main code base. The build is run again on the master but should almost never fail, since it didn't fail on the branch.

The forking workflow

The forking workflow is very popular among open source communities. It is presented in the following diagram:

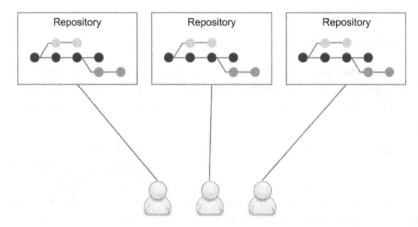

Figure 4.12 – The forking workflow

Each developer has their own server-side repository. It may or may not be the official repository, but technically, each repository is exactly the same.

Forking means literally creating a new repository from another repository. Developers push to their own repositories, and when they want to integrate code, they create a pull request to the other repository.

The main advantage of the forking workflow is that the integration is not necessarily via a central repository. It also helps with ownership because it allows the acceptance of pull requests from others without giving them write access.

In the case of requirement-oriented commercial projects, a team usually works on one product and therefore has a central repository, so this model boils down to having a branching workflow with good ownership assignment; for example, only project leads can merge pull requests into the central repository.

Adopting continuous integration

We have described different development workflows, but *how do they influence the continuous integration configuration?*

Branching strategies

Each development workflow implies a different continuous integration approach:

- **Trunk-based workflow**: This implies constantly struggling against the broken pipeline. If everyone commits to the main code base, the pipeline often fails. In this case, the old continuous integration rule says, *If the build is broken, the development team stops whatever they are doing and fixes the problem immediately.*

- **Branching workflow**: This solves the broken trunk issue but introduces another one: if everyone develops in their own branches, *where is the integration?* A feature usually takes weeks or months to develop, and for all this time, the branch is not integrated into the main code. Therefore, it cannot really be called continuous integration – not to mention that there is a constant need for merging and resolving conflicts.

- **Forking workflow**: This implies managing the continuous integration process by every repository owner, which isn't usually a problem. It does share, however, the same issues as the branching workflow.

There is no silver bullet, and different organizations choose different strategies. The solution that is the closest to perfection uses the technique of the branching workflow and the philosophy of the trunk-based workflow. In other words, we can create very small branches and integrate them frequently into the master. This seems to take the best aspects of both. However, it requires either having tiny features or using feature toggles. Since the concept of feature toggles fits very well into continuous integration and continuous delivery, let's take a moment to explore it.

Feature toggles

Feature toggles is a technique that is an alternative to maintaining multiple source code branches, such that the feature can be tested before it is completed and ready for release. It is used to disable the feature for users but enable it for developers while testing. Feature toggles are essentially variables used in conditional statements.

The simplest implementation of feature toggles are flags and the `if` statements. A development using feature toggles, as opposed to a feature branching development, appears as follows:

1. A new feature has to be implemented.
2. Create a new flag or a configuration property – `feature_toggle` (instead of the `feature` branch).

3. All feature-related code is added inside the `if` statement (instead of committing to the `feature` branch), such as the following:

```
if (feature_toggle) {
    // do something
}
```

4. During the feature development, the following takes place:

    - Coding is done in the master with `feature_toggle = true` (instead of coding in the feature branch).

    - The release is done from the master with `feature_toggle = false`.

5. When the feature development is completed, all `if` statements are removed and `feature_toggle` is removed from the configuration (instead of merging `feature` to the master and removing the `feature` branch).

The benefit of feature toggles is that all development is done in the trunk, which facilitates real continuous integration and mitigates problems with merging the code.

Jenkins multi-branch

If you decide to use branches in any form, either the long-feature branches or the recommended short-lived branches, it is convenient to know that the code is healthy before merging it into the master. This approach results in always keeping the main code base green, and luckily, there is an easy way to do it with Jenkins.

In order to use multi-branch in our calculator project, let's proceed with the following steps:

1. Open the main Jenkins page.

2. Click on **New Item**.

3. Enter `calculator-branches` as the item name, select **Multibranch Pipeline**, and click on **OK**.

4. In the **Branch Sources** section, click on **Add source**, and select **Git**.

5. Enter the repository address in the **Project Repository** field:

Figure 4.13 – The multi-branch pipeline configuration

6. Tick **Periodically if not otherwise run** and set **1 minute** as the interval.

7. Click on **Save**.

Every minute, this configuration checks whether there were any branches added (or removed) and creates (or deletes) the dedicated pipeline defined by `Jenkinsfile`.

We can create a new branch and see how it works. Let's create a new branch called `feature` and push it into the repository:

```
$ git checkout -b feature
$ git push origin feature
```

After a moment, you should see a new branch pipeline automatically created and run:

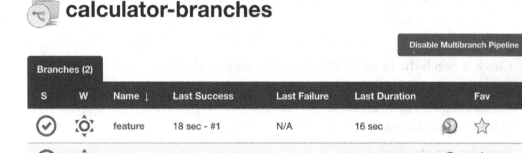

Figure 4.14 – The multi-branch pipeline build

Now, before merging the feature branch to the master, we can check whether it's green. This approach should never break the master build.

A very similar approach is to build a pipeline per pull request instead of a pipeline per branch, which gives the same result – the main code base is always healthy.

Non-technical requirements

Last but not least, continuous integration is not all about technology. On the contrary, technology comes second. James Shore, in his *Continuous Integration on a Dollar a Day* article, described how to set up the continuous integration process without any additional software. All he used was a rubber chicken and a bell. The idea is to make the team work in one room and set up a separate computer with an empty chair. Put the rubber chicken and the bell in front of that computer. Now, when you plan to check in the code, take the rubber chicken, check in the code, go to the empty computer, check out the fresh code, run all tests there, and if everything passes, put back the rubber chicken, and ring the bell so that everyone knows that something has been added to the repository.

> **Information**
> *Continuous Integration on a Dollar a Day* by *James Shore* can be found at http://www.jamesshore.com/v2/blog/2006/ continuous-integration-on-a-dollar-a-day.

The idea is a little oversimplified, and automated tools are useful; however, the main message is that without each team member's engagement, even the best tools won't help. In his book, Jez Humble outlines the prerequisites for continuous integration:

- **Check in regularly**: To quote Mike Roberts, *continuously is more often than you think*; the minimum is once a day.

- **Create comprehensive unit tests**: It's not only about high test coverage; it's possible to have no assertions and still keep 100% coverage.

- **Keep the process quick**: Continuous integration must take a short time, preferably under 5 minutes. 10 minutes is already a lot.

- **Monitor the builds**: This can be a shared responsibility, or you can adapt the **build master** role that rotates weekly.

Summary

In this chapter, we covered all aspects of the continuous integration pipeline, which is always the first step for continuous delivery. Here are the key takeaways:

- The pipeline provides a general mechanism for organizing any automation processes; however, the most common use cases are continuous integration and continuous delivery.

- Jenkins accepts different ways of defining pipelines, but the recommended one is the declarative syntax.

- The commit pipeline is the most basic continuous integration process, and as its name suggests, it should be run after every commit to the repository.

- The pipeline definition should be stored in the repository as a `Jenkinsfile` file.

- The commit pipeline can be extended with the code quality stages.

- No matter what the project build tool, Jenkins commands should always be consistent with local development commands.

- Jenkins offers a wide range of triggers and notifications.

- The development workflow should be carefully chosen inside a team or organization because it affects the continuous integration process and defines the way code is developed.

In the next chapter, we will focus on the next phase of the continuous delivery process – automated acceptance testing. This can be considered the most important and, in many cases, the most difficult step to implement. We will explore the idea of acceptance testing and a sample implementation using Docker.

Exercises

You've learned a lot about how to configure the continuous integration process. Since *practice makes perfect*, I recommend doing the following exercises:

- Create a Python program that multiplies two numbers passed as command-line parameters. Add unit tests and publish the project on GitHub:

    I.   Create two files: `calculator.py` and `test_calculator.py`.

    II.   You can use the `unittest` library at `https://docs.python.org/3/library/unittest.html`.

    III. Run the program and the unit test.

- Build the continuous integration pipeline for the Python calculator project:

    I.   Use `Jenkinsfile` to specify the pipeline.

    II.   Configure the trigger so that the pipeline runs automatically in case of any commits to the repository.

    III. The pipeline doesn't need the `Compile` step since Python is an interpretable language.

    IV.   Run the pipeline and observe the results.

    V.   Try to commit the code that breaks the pipeline build and observe how it is visualized in Jenkins.

Questions

To verify the knowledge acquired from this chapter, please answer the following questions:

1. What is a pipeline?
2. What is the difference between a *stage* and a *step* in the pipeline?
3. What is the `post` section in the Jenkins pipeline?
4. What are the three most fundamental stages of the commit pipeline?
5. What is `Jenkinsfile`?

6. What is the purpose of the code coverage stage?

7. What is the difference between the following Jenkins triggers – external and polling SCM?

8. What are the most common Jenkins notification methods? Name at least three.

9. What are the three most common development workflows?

10. What is a feature toggle?

Further reading

To read more about the continuous integration topic, please refer to the following resources:

- *Continuous Delivery, Jez Humble and David Farley*: `https://continuousdelivery.com/`

- *Continuous Integration: Improving Software Quality and Reducing Risk, Andrew Glover, Steve Matyas, and Paul M. Duvall*: `https://www.oreilly.com/library/view/continuous-integration-improving/9780321336385/`

5
Automated Acceptance Testing

We've configured the commit phase of the **continuous delivery** (**CD**) process and it's now time to address the acceptance testing phase, which is usually the most challenging part. By gradually extending the pipeline, we will see different aspects of a well-executed acceptance testing automation.

This chapter covers the following topics:

- Introducing acceptance testing
- Installing and using the Docker Registry
- Acceptance tests in the Jenkins pipeline
- Writing acceptance tests

Technical requirements

To complete this chapter, you'll need the following software:

- Jenkins
- Docker
- The **Java Development Kit** (**JDK**) 8+

All examples and solutions to the exercises can be found at `https://github.com/PacktPublishing/Continuous-Delivery-With-Docker-and-Jenkins-3rd-Edition/tree/main/Chapter05`.

Code in Action videos for this chapter can be viewed at `https://bit.ly/3Ki1aIm`.

Introducing acceptance testing

Acceptance testing is a step performed to determine whether the business requirements or contracts are met. It involves black-box testing against a complete system from a user perspective, and a positive result means acceptance of the software delivery. Sometimes also called **user acceptance testing** (**UAT**) or end-user testing, it is a phase of the development process where software meets a *real-world* audience.

Many projects rely on manual steps performed by **quality assurers** (**QAs**) or users to verify the **functional** and **non-functional requirements** (**FRs** and **NFRs**), but still, it's way more reasonable to run them as programmed repeatable operations.

Automated acceptance tests, however, can be considered difficult due to their specifics, as outlined here:

- **User-facing**: They need to be written together with a user, which requires an understanding between two worlds—technical and non-technical.

- **Dependencies integration**: The tested application should be run together with its dependencies in order to check that the system as a whole works properly.

- **Staging environment**: The staging (testing) environment needs to be identical to the production one so as to ensure the same functional and non-functional behavior.

- **Application identity**: Applications should be built only once, and the same binary should be transferred to production. This eliminates the risk of different building environments.

- **Relevance and consequences**: If the acceptance test passes, it should be clear that the application is ready for release from the user's perspective.

We address all these difficulties in different sections of this chapter. Application identity can be achieved by building the Docker image only once and using Docker Registry for its storage and versioning. Creating tests in a user-facing manner is explained in the *Writing acceptance tests* section, and the environment identity is addressed by the Docker tool itself and can also be improved by other tools described in the next chapters.

> **Information**
>
> Acceptance testing can have multiple meanings; in this book, we treat acceptance testing as a complete integration test suite from a user perspective, excluding NFRs such as performance, load, and recovery.

Since we understand the goal and meaning of acceptance testing, let's describe the first aspect we need—the **Docker Registry**.

Installing and using the Docker Registry

The Docker Registry is a store for Docker images. To be precise, it is a stateless server application that allows the images to be published (pushed) and later retrieved (pulled). In *Chapter 2*, *Introducing Docker*, we already saw an example of the Registry when running the official Docker images, such as `hello-world`. We pulled the images from Docker Hub, which is an official cloud-based Docker Registry. Having a separate server to store, load, and search software packages is a more general concept called the software repository or, in even more general terms, the artifact repository. Let's look closer at this idea.

The artifact repository

While the source control management stores the source code, the artifact repository is dedicated to storing software binary artifacts, such as compiled libraries or components, later used to build a complete application. *Why do we need to store binaries on a separate server using a separate tool?* Here's why:

- **File size**: Artifact files can be large, so the systems need to be optimized for their download and upload.

- **Versions**: Each uploaded artifact needs to have a version that makes it easy to browse and use. Not all versions, however, have to be stored forever; for example, if there was a bug detected, we may not be interested in the related artifact and remove it.

- **Revision mapping**: Each artifact should point to exactly one revision of the source control and, what's more, the binary creation process should be repeatable.

- **Packages**: Artifacts are stored in a compiled and compressed form so that these time-consuming steps don't need to be repeated.

- **Access control**: Users can be restricted differently in terms of access to the source code and artifact binary.

- **Clients**: Users of the artifact repository can be developers outside the team or organization who want to use the library via its public **application programming interface (API)**.

- **Use cases**: Artifact binaries are used to guarantee that exactly the same build version is deployed to every environment to ease the rollback procedure in case of failure.

> **Information**
>
> The most popular artifact repositories are **JFrog Artifactory** and **Sonatype Nexus**.

The artifact repository plays a special role in the CD process because it guarantees that the same binary is used throughout all pipeline steps.

Let's look at the following diagram to understand how it works:

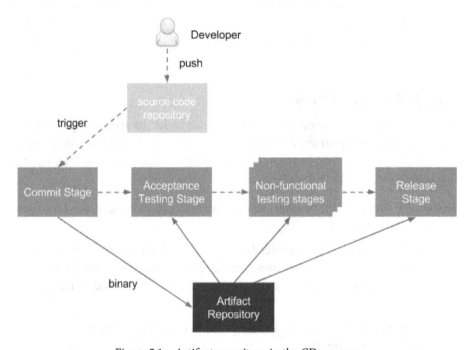

Figure 5.1 – Artifact repository in the CD process

The **developer** pushes a change to the **source code repository**, which triggers the pipeline build. As the last step of the **commit stage**, a binary is created and stored in the **artifact repository**. Afterward, during all other stages of the delivery process, the same binary is (pulled and) used.

> **Information**
>
> The binary is often called the **release candidate**, and the process of moving the binary to the next stage is called **promotion**.

Depending on the programming language and technologies, the binary formats can differ. For example, in the case of Java, **Java ARchive (JAR)** files are usually stored and, in the case of Ruby, gem files. We work with Docker, so we will store Docker images as artifacts, and the tool to store Docker images is called the **Docker Registry**.

> **Information**
>
> Some teams maintain two repositories at the same time; the artifact repository for JAR files and the Docker Registry for Docker images. While it may be useful during the first phase of the Docker introduction, there is no good reason to maintain both forever.

Installing a Docker Registry

First, we need to install a Docker Registry. There are a number of options available, but all of them fall into two categories: a cloud-based Docker Registry and a self-hosted Docker Registry. Let's dig into them.

Cloud-based Docker Registry

The benefit of using a cloud-based service is that you don't need to install or maintain anything on your own. There are a number of cloud offerings available; however, Docker Hub is by far the most popular. That is why we will use it throughout this book.

Docker Hub

Docker Hub provides a Docker Registry service and other related features, such as building images, testing them, and pulling code directly from the code repository. Docker Hub is cloud-hosted, so it does not really need any installation process. All you need to do is create a Docker Hub account, as follows:

1. Open `https://hub.docker.com/` in a browser.
2. In **Sign Up**, fill in the password, email address, and Docker **identifier (ID)**.
3. After receiving an email and clicking the activation link, an account is created.

Docker Hub is definitely the simplest option to start with, and it allows the storing of both private and public images.

Docker Hub alternatives

There are more cloud offerings worth mentioning. First of all, each of the following three main cloud platforms offers its own Docker Registry:

- Amazon **Elastic Container Registry (ECR)**
- Google Artifact Registry
- Azure Container Registry

Other widely used solutions include the following:

- Quay Container Registry
- JFrog Artifactory
- GitLab Container Registry

All of the mentioned registries implement the same Docker Registry protocol, so the good news is that no matter which you choose, the commands used are exactly the same.

Self-hosted Docker Registry

Cloud solutions may not always be acceptable. They are not free for enterprises and, what's even more important, a lot of companies have policies not to store their software outside their own network. In this case, the only option is to install a self-hosted Docker Registry.

The Docker Registry installation process is quick and simple, but making it secure and available in public requires setting up access restrictions and the domain certificate. This is why we split this section into three parts, as follows:

- Installing the Docker Registry application
- Adding a domain certificate
- Adding an access restriction

Let's have a look at each part.

Installing the Docker Registry application

The Docker Registry is available as a Docker image. To start this, we can run the following command:

```
$ docker run -d -p 5000:5000 --restart=always --name registry
registry:2
```

> **Tip**
>
> By default, the registry data is stored as a Docker volume in the default host filesystem's directory. To change it, you can add `-v <host_directory>:/var/lib/registry`. Another alternative is to use a volume container.

The command starts the registry and makes it accessible through port `5000`. The `registry` container is started from the registry image (version 2). The `--restart=always` option causes the container to automatically restart whenever it's down.

> **Tip**
>
> Consider setting up a load balancer and starting a few Docker Registry containers in case of a large number of users. Note that, in such a case, they need to share the storage or have a synchronization mechanism in place.

Adding a domain certificate

If the registry is run on the localhost, then everything works fine and no other installation steps are required. However, in most cases, we want to have a dedicated server for the registry so that the images are widely available. In that case, Docker requires the securing of the registry with **Secure Sockets Layer/Transport Layer Security (SSL/TLS)**. The process is very similar to the public web server configuration and, similarly, it's highly recommended that you have the certificate signed by a **certificate authority (CA)**. If obtaining a CA-signed certificate is not an option, we can self-sign a certificate or use the `--insecure-registry` flag.

> **Information**
>
> You can read about creating and using self-signed certificates at `https://docs.docker.com/registry/insecure/#use-self-signed-certificates`.

Once the certificates are either signed by the CA or self-signed, we can move `domain.crt` and `domain.key` to the `certs` directory and start the registry, which listens on the default **HyperText Transfer Protocol Secure (HTTPS)** port, as follows:

```
$ docker run -d -p 443:443 --restart=always --name registry
-v `pwd`/certs:/certs -e REGISTRY_HTTP_ADDR=0.0.0.0:443 -e
REGISTRY_HTTP_TLS_CERTIFICATE=/certs/domain.crt -e REGISTRY_
HTTP_TLS_KEY=/certs/domain.key registry:2
```

Using the --insecure-registry flag is not recommended since it provides no proper CA verification.

> **Information**
> Read more about setting up Docker registries and making them secure in the official Docker documentation at https://docs.docker.com/registry/deploying/.

Adding an access restriction

Unless we use the registry inside a highly secure private network, we should configure authentication.

The simplest way to do this is to create a user with a password using the htpasswd tool from the registry image, as follows:

```
$ mkdir auth
$ docker run --entrypoint htpasswd httpd:2 -Bbn <username>
<password> > auth/htpasswd
```

The command runs the htpasswd tool to create an auth/htpasswd file (with one user inside). Then, we can run the registry with that one user authorized to access it, like this:

```
$ docker run -d -p 443:443 --restart=always --name registry
-v `pwd`/auth:/auth -e "REGISTRY_AUTH=htpasswd" -e "REGISTRY_
AUTH_HTPASSWD_REALM=Registry Realm" -e REGISTRY_AUTH_HTPASSWD_
PATH=/auth/htpasswd -v `pwd`/certs:/certs -e REGISTRY_
HTTP_ADDR=0.0.0.0:443 -e REGISTRY_HTTP_TLS_CERTIFICATE=/
certs/domain.crt -e REGISTRY_HTTP_TLS_KEY=/certs/domain.key
registry:2
```

The command, in addition to setting the certificates, creates an access restriction limited to the users specified in the auth/passwords file.

As a result, before using the registry, a client needs to specify the username and password.

> **Important Note**
> Access restriction doesn't work in the case of the --insecure-registry flag.

Using the Docker Registry

When our registry is configured, we can show how to work with it in three stages, as follows:

- Building an image
- Pushing the image into the registry
- Pulling the image from the registry

Building an image

Let's use the example from *Chapter 2*, *Introducing Docker*, and build an image with Ubuntu and the Python interpreter installed. In a new directory, we need to create a Dockerfile, as follows:

```
FROM ubuntu:20.04
RUN apt-get update && \
    apt-get install -y python
```

Now, we can build the image with the following command:

```
$ docker build -t ubuntu_with_python .
```

After the image is built, we can push it into the Docker Registry.

Pushing the image into the registry

In order to push the created image, we need to tag it according to the naming convention, like this:

```
<registry_address>/<image_name>:<tag>
```

The registry_address value can be either of the following:

- A username in the case of Docker Hub
- A domain name or **Internet Protocol** (**IP**) address with a port for a private registry (for example, localhost:5000)

> **Information**
> In most cases, <tag> is in the form of the image/application version.

Let's tag the image to use Docker Hub, as follows:

```
$ docker tag ubuntu_with_python leszko/ubuntu_with_python:1
```

Remember to use your Docker Hub username instead of `leszko`.

> **Tip**
> We could have also tagged the image in the `build` command, like this:
> `docker build -t leszko/ubuntu_with_python:1`.

If the repository has access restriction configured, we need to authorize it first, like this:

```
$ docker login --username <username> --password <password>
```

> **Information**
> If you use a Docker Registry other than Docker Hub, then you also need to add a **Uniform Resource Locator** (**URL**) to the `login` command—for example, `docker login quay.io`.

Now, we can store the image in the registry using the `push` command, as follows:

```
$ docker push leszko/ubuntu_with_python:1
```

Note that there is no need to specify the registry address because Docker uses the naming convention to resolve it. The image is stored, and we can check it using the Docker Hub web interface available at `https://hub.docker.com`.

Pulling the image from the registry

To demonstrate how the registry works, we can remove the image locally and retrieve it from the registry, like this:

```
$ docker rmi ubuntu_with_python leszko/ubuntu_with_python:1
```

We can see that the image has been removed using the `docker images` command. Then, let's retrieve the image back from the registry by executing the following code:

```
$ docker pull leszko/ubuntu_with_python:1
```

> **Tip**
>
> If you use a free Docker Hub account, you may need to change the `ubuntu_with_python` repository to `public` before pulling it.

We can confirm that the image is back with the `docker images` command.

When we have the registry configured and understand how it works, we can see how to use it inside the CD pipeline and build the acceptance testing stage.

Acceptance tests in the Jenkins pipeline

We already understand the idea behind acceptance testing and know how to configure the Docker Registry, so we are ready for its first implementation inside the Jenkins pipeline.

Let's look at the following diagram, which presents the process we will use:

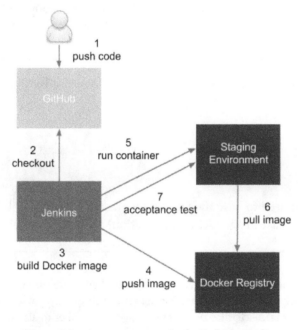

Figure 5.2 – Acceptance tests in the Jenkins pipeline

The process goes like this:

1. The developer pushes a code change to GitHub.
2. Jenkins detects the change, triggers the build, and checks out the current code.
3. Jenkins executes the commit phase and builds the Docker image.
4. Jenkins pushes the image to the **Docker Registry**.
5. Jenkins runs the Docker container in the staging environment.
6. The Docker host on the staging environment needs to pull the image from the Docker Registry.
7. Jenkins runs the acceptance test suite against the application running in the staging environment.

> **Information**
>
> For the sake of simplicity, we will run the Docker container locally (and not on a separate staging server). In order to run it remotely, we need to use the -H option or configure the DOCKER_HOST environment variable.

Let's continue the pipeline we started in *Chapter 4, Continuous Integration Pipeline*, and add three more stages, as follows:

- Docker build
- Docker push
- Acceptance test

Keep in mind that you need to have the Docker tool installed on the Jenkins executor (agent or master, in the case of agentless configuration) so that it can build Docker images.

> **Tip**
>
> If you use dynamically provisioned Docker agents, then make sure you use the **Docker-in-Docker (DinD)** solution. For the sake of simplicity, you can use the leszko/jenkins-docker-slave image. Remember to also mark the privileged option in the Docker agent configuration.

The Docker build stage

We would like to run the calculator project as a Docker container, so we need to create a Dockerfile and add the `Docker build` stage to the Jenkinsfile.

Adding a Dockerfile

Let's create a Dockerfile in the root directory of the calculator project, as follows:

```
FROM openjdk:11-jre
COPY build/libs/calculator-0.0.1-SNAPSHOT.jar app.jar
ENTRYPOINT ["java", "-jar", "app.jar"]
```

> **Information**
>
> The default build directory for Gradle is `build/libs/`, and `calculator-0.0.1-SNAPSHOT.jar` is the complete application packaged into one JAR file. Note that Gradle automatically versioned the application using the `0.0.1-SNAPSHOT` Maven-style version.

The Dockerfile uses a base image that contains the **Java Runtime Environment 11 (JRE 11)** (`openjdk:11-jre`). It also copies the application JAR (created by Gradle) and runs it. Let's now check whether the application builds and runs by executing the following code:

```
$ ./gradlew build
$ docker build -t calculator .
$ docker run -p 8080:8080 --name calculator calculator
```

Using the preceding commands, we've built the application, built the Docker image, and run the Docker container. After a while, we should be able to open the browser at `http://localhost:8080/sum?a=1&b=2` and see 3 as a result.

We can stop the container and push the Dockerfile to the GitHub repository, like this:

```
$ git add Dockerfile
$ git commit -m "Add Dockerfile"
$ git push
```

Adding the Docker build to the pipeline

The final step we need to perform is to add the Docker build stage to the Jenkinsfile. Usually, the JAR packaging is also declared as a separate Package stage, as illustrated in the following code snippet:

```
stage("Package") {
    steps {
        sh "./gradlew build"
    }
}

stage("Docker build") {
    steps {
        sh "docker build -t leszko/calculator ."
    }
}
```

> **Information**
>
> We don't explicitly version the image, but each image has a unique hash ID. We will cover explicit versioning in the following chapters.

Note that we used the Docker Registry name in the image tag. There is no need to have the image tagged twice as calculator and leszko/calculator.

When we commit and push the Jenkinsfile, the pipeline build should start automatically and we should see all boxes in green. This means that the Docker image has been built successfully.

> **Tip**
>
> If you see a failure in the Docker build stage, then most probably, your Jenkins executor doesn't have access to the Docker daemon. In case you use the Jenkins master as the executor, make sure that the jenkins user is added to the docker user group. In case you use Jenkins agents, make sure they have access to the Docker daemon.

The Docker push stage

When the image is ready, we can store it in the registry. The `Docker` push stage is very simple. It's enough to add the following code to the Jenkinsfile:

```
stage("Docker push") {
    steps {
        sh "docker push leszko/calculator"
    }
}
```

> **Information**
>
> If the Docker Registry has access restricted, first, we need to log in using the `docker login` command. Needless to say, the credentials must be well secured—for example, using a dedicated credential store, as described on the official Docker page at `https://docs.docker.com/engine/reference/commandline/login/#credentials-store`.

As always, pushing changes to the GitHub repository triggers Jenkins to start the build and, after a while, we should have the image automatically stored in the registry.

The acceptance testing stage

To perform acceptance testing, we first need to deploy the application to the staging environment and then run the acceptance test suite against it.

Adding a staging deployment to the pipeline

Let's add a stage to run the `calculator` container, as follows:

```
stage("Deploy to staging") {
    steps {
        sh "docker run -d --rm -p 8765:8080 --name calculator leszko/calculator"
    }
}
```

After running this stage, the `calculator` container is running as a daemon, publishing its port as `8765`, and being removed automatically when stopped.

Finally, we are ready to add the acceptance test to our Jenkins pipeline.

Adding an acceptance test to the pipeline

Acceptance testing usually requires running a dedicated black-box test suite that checks the behavior of the system. We will cover it in the *Writing acceptance tests* section. At the moment, for the sake of simplicity, let's perform acceptance testing simply by calling the web service endpoint with the curl tool and checking the result using the test command.

In the root directory of the project, let's create an acceptance_test.sh file, as follows:

```
#!/bin/bash
test $(curl localhost:8765/sum?a=1\&b=2) -eq 3
```

We call the sum endpoint with the a=1 and b=2 parameters and expect to receive 3 in response.

Then, an Acceptance test stage can be added, as follows:

```
stage("Acceptance test") {
        steps {
                sleep 60
                sh "chmod +x acceptance_test.sh && ./acceptance_test.
sh"
        }
}
```

Since the docker run -d command is asynchronous, we need to wait, using the sleep operation to make sure the service is already running.

> **Information**
>
> There is no good way to check whether the service is already running. An alternative to sleeping could be a script checking every second to see whether the service has already started.

At this point, our pipeline has already performed the automated acceptance tests. One last thing we should never forget about is to add a cleanup stage.

Adding a cleanup stage environment

As the final stage of acceptance testing, we can add the staging environment cleanup. The best place to do this is in the `post` section, to make sure it executes even in case of failure. Here's the code we need to execute:

```
post {
    always {
        sh "docker stop calculator"
    }
}
```

This statement makes sure that the `calculator` container is no longer running on the Docker host.

Writing acceptance tests

So far, we used the `curl` command to perform a suite of acceptance tests. That is, obviously, a considerable simplification. Technically speaking, if we write a **REpresentational State Transfer** (**REST**) web service, we could write all black-box tests as a big script with a number of `curl` calls. However, this solution would be very difficult to read, understand, and maintain. What's more, the script would be completely incomprehensible to non-technical, business-related users. *How do we address this issue and create tests with a good structure that are readable by users and meet their fundamental goal: automatically checking that the system is as expected?* I will answer this question throughout this section.

Writing user-facing tests

Acceptance tests are written with users and should be comprehensible to users. This is why the choice of a method for writing them depends on who the customer is.

For example, imagine a purely technical person. If you write a web service that optimizes database storage and your system is used only by other systems and read-only by other developers, your tests can be expressed in the same way as unit tests. As a rule, the test is good if understood by both developers and users.

In real life, most software is written to deliver a specific business value, and that business value is defined by non-developers. Therefore, we need a common language to collaborate. On one side, there is the business, which understands what is needed but not how to do it; on the other side, the development team knows how but doesn't know what. Luckily, there are a number of frameworks that helps to connect these two worlds, such as **Cucumber**, **FitNesse**, **JBehave**, and **Capybara**. They differ from each other, and each of them may be a subject for a separate book; however, the general idea of writing acceptance tests is the same and is shown in the following diagram:

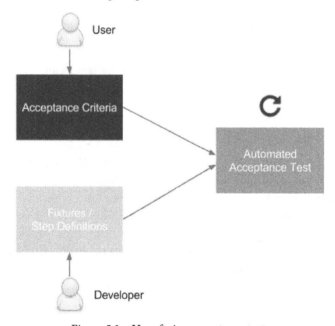

Figure 5.3 – User-facing acceptance tests

The **acceptance criteria** are written by users (or a product owner as their representative), with the help of developers. They are usually written in the form of the following scenarios:

```
Given I have two numbers: 1 and 2
When the calculator sums them
Then I receive 3 as a result
```

Developers write the testing implementation, called **fixtures** or **step definitions**, that integrates the human-friendly **domain-specific language** (**DSL**) specification with the programming language. As a result, we have an automated test that can be easily integrated into the CD pipeline.

Needless to add, writing acceptance tests is a continuous Agile process, not a Waterfall one. It requires constant collaboration, during which the test specifications are improved and maintained by both developers and the business.

Information

In the case of an application with a **user interface (UI)**, it can be tempting to perform the acceptance test directly through the interface (for example, by recording Selenium scripts). However, this approach, when not done properly, can lead to tests that are slow and tightly coupled to the interface layer.

Let's see how writing acceptance tests looks in practice and how to bind them to the CD pipeline.

Using the acceptance testing framework

Let's use the Cucumber framework and create an acceptance test for the calculator project. As previously described, we will do this in three stages, as follows:

1. Creating acceptance criteria

2. Creating step definitions

3. Running an automated acceptance test

Creating acceptance criteria

Let's put the business specification in `src/test/resources/feature/` `calculator.feature`, as follows:

```
Feature: Calculator
  Scenario: Sum two numbers
    Given I have two numbers: 1 and 2
    When the calculator sums them
    Then I receive 3 as a result
```

This file should be created by users with the help of developers. Note that it is written in a way that non-technical people can understand.

Creating step definitions

The next step is to create Java bindings so that the feature specification will be executable. In order to do this, we create a new file, src/test/java/acceptance/ StepDefinitions.java, as follows:

```java
package acceptance;

import io.cucumber.java.en.Given;
import io.cucumber.java.en.Then;
import io.cucumber.java.en.When;
import org.springframework.web.client.RestTemplate;

import static org.junit.Assert.assertEquals;

/** Steps definitions for calculator.feature */
public class StepDefinitions {
    private String server = System.getProperty("calculator.url");

    private RestTemplate restTemplate = new RestTemplate();

    private String a;
    private String b;
    private String result;

    @Given("^I have two numbers: (.*) and (.*)$")
    public void i_have_two_numbers(String a, String b) throws Throwable {
        this.a = a;
        this.b = b;
    }

    @When("^the calculator sums them$")
    public void the_calculator_sums_them() throws Throwable {
        String url = String.format("%s/sum?a=%s&b=%s", server, a, b);
        result = restTemplate.getForObject(url, String.class);
    }
```

```
@Then("^I receive (.*) as a result$")
    public void i_receive_as_a_result(String expectedResult)
throws Throwable {
        assertEquals(expectedResult, result);
    }
}
```

Each line (`Given`, `When`, and `Then`) from the feature specification file is matched by **regular expressions** (**regexes**) with the corresponding method in the Java code. Wildcards (`.*`) are passed as parameters. Note that the server address is passed as the `calculator.url` Java property. The method performs the following actions:

- `i_have_two_numbers`: Saves parameters as fields
- `the_calculator_sums_them`: Calls the remote calculator service and stores the result in a field
- `i_receive_as_a_result`: Asserts that the result is as expected

Running an automated acceptance test

To run an automated test, we need to make a few configurations, as follows:

1. Add the Java Cucumber libraries. In the `build.gradle` file, add the following code to the `dependencies` section:

    ```
    testImplementation("io.cucumber:cucumber-
    java:7.2.0")
    testImplementation("io.cucumber:cucumber-
    junit:7.2.0")
    ```

2. Add the Gradle target. In the same file, add the following code:

    ```
    tasks.register('acceptanceTest', Test) {
    include '**/acceptance/**'
    systemProperties System.getProperties()
    }

    test {
    useJUnitPlatform()
    exclude '**/acceptance/**'
    }
    ```

This splits the tests into unit tests (run with `./gradlew test`) and acceptance tests (run with `./gradlew acceptanceTest`).

3. Add a JUnit Test Runner, add a new file, `src/test/java/acceptance/AcceptanceTest.java`, as follows:

```
package acceptance;

import io.cucumber.junit.CucumberOptions;
import io.cucumber.junit.Cucumber;
import org.junit.runner.RunWith;

/** Acceptance Test */
@RunWith(Cucumber.class)
@CucumberOptions(features = "classpath:feature")
public class AcceptanceTest { }
```

This is the entry point to the acceptance test suite.

After this configuration, if the server is running on the localhost, we can test it by executing the following code:

```
$ ./gradlew acceptanceTest \
 -Dcalculator.url=http://localhost:8765
```

Obviously, we can add this command instead of `acceptance_test.sh`. This would make the Cucumber acceptance test run in the Jenkins pipeline.

Acceptance test-driven development

Acceptance tests, as with most aspects of the CD process, are less about technology and more about people. The test quality depends, of course, on the engagement of users and developers, but also, what is maybe less intuitive is the time when the tests are created.

The last question to ask is this: *During which phase of the software development life cycle should the acceptance tests be prepared?* Or, to rephrase it: *Should we create acceptance tests before or after writing the code?*

Technically speaking, the result is the same; the code is well covered with both unit and acceptance tests. However, it's tempting to consider writing tests first. The idea of **test-driven development** (**TDD**) can be well adapted for acceptance testing. If unit tests are written before the code, the resulting code is cleaner and better structured. Analogously, if acceptance tests are written before the system feature, the resulting feature corresponds better to the customer's requirements.

This process, often called acceptance TDD, is presented in the following diagram:

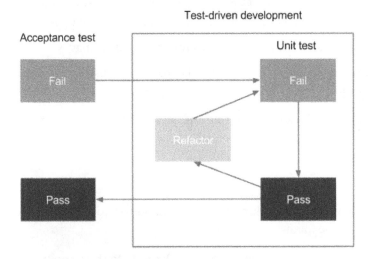

Figure 5.4 – Acceptance TDD

Users (with developers) write the acceptance criteria specification in the human-friendly DSL format. Developers write the fixtures, and the tests fail. Then, feature development starts using the TDD methodology internally. Once the feature is completed, the acceptance test should pass, and this is a sign that the feature is completed.

A very good practice is to attach the Cucumber feature specification to the request ticket in the issue-tracking tool (for example, JIRA) so that the feature would always be requested together with its acceptance test. Some development teams take an even more radical approach and refuse to start the development process if no acceptance tests are prepared. There is a lot of sense in that. After all, how can you develop something that the client can't test?

Summary

In this chapter, you learned how to build a complete and functional acceptance test stage, which is an essential part of the CD process. Here are the key takeaways:

- Acceptance tests can be difficult to create because they combine technical challenges (application dependencies; setting up the environment) with personal challenges (developer/business collaboration).

- Acceptance testing frameworks provide a way to write tests in a human-friendly language that makes them comprehensible to non-technical people.

- The Docker Registry is an artifact repository for Docker images.

- The Docker Registry fits well with the CD process because it provides a way to use exactly the same Docker image throughout the stages and environments.

In the next chapter, we will cover clustering and service dependencies, which is the next step toward creating a complete CD pipeline.

Exercises

We covered a lot of new material throughout this chapter, so to aid your understanding, I recommend doing the following exercises:

1. Create a Ruby-based web service, `book-library`, to store books.

   The acceptance criteria are delivered in the form of the following Cucumber feature:

   ```
   Scenario: Store book in the library
       Given Book "The Lord of the Rings" by "J.R.R. Tolkien"
   with ISBN number "0395974682"
       When I store the book in library
       Then I am able to retrieve the book by the ISBN number
   ```

   Proceed as follows:

   I.   Write step definitions for the Cucumber test.

   II. Write the web service (the simplest way is to use the Sinatra framework (http://www.sinatrarb.com/), but you can also use Ruby on Rails).

   III. The book should have the following attributes: `name`, `author`, and **International Standard Book Number** (`ISBN`).

IV. The web service should have the following endpoints:

- POST /books to add a book

- GET /books/<isbn> to retrieve the book

V. The data can be stored in the memory.

VI. At the end, check that the acceptance test is green.

2. Add book-library as a Docker image to the Docker Registry by doing the following:

    I.   Create an account on Docker Hub.

    II. Create a Dockerfile for the application.

    III. Build the Docker image and tag it according to the naming convention.

    IV. Push the image to Docker Hub.

3. Create a Jenkins pipeline to build the Docker image, push it to the Docker Registry, and perform acceptance testing by doing the following:

    I.   Create a Docker build stage.

    II. Create Docker login and Docker push stages.

    III. Add an Acceptance test stage to the pipeline.

    IV. Run the pipeline and observe the result.

Questions

To verify the knowledge acquired from this chapter, please answer the following questions:

1. What is the Docker Registry?

2. What is Docker Hub?

3. What is the convention for naming Docker images (later pushed to the Docker Registry)?

4. What is the staging environment?

5. Which Docker commands would you use to build an image and push it into Docker Hub?

6. What is the main purpose of acceptance testing frameworks such as Cucumber and FitNesse?

7. What are the three main parts of a Cucumber test?

8. What is acceptance TDD?

Further reading

To learn more about Docker Registry, acceptance testing, and Cucumber, please refer to the following resources:

- **Docker Registry documentation**: `https://docs.docker.com/registry/`

- *Jez Humble, David Farley—Continuous Delivery*: `https://continuousdelivery.com/`

- **Cucumber framework**: `https://cucumber.io/`

6
Clustering with Kubernetes

So far, in this book, we have covered the fundamental aspects of the acceptance testing process. In this chapter, we will see how to change the Docker environment from a single Docker host into a cluster of machines and how to change an independent application into a system composed of multiple applications.

This chapter covers the following topics:

- Server clustering
- Introducing Kubernetes
- Kubernetes installation
- Using Kubernetes
- Advanced Kubernetes
- Application dependencies
- Alternative cluster management systems

Technical requirements

To follow along with the instructions in this chapter, you'll need the following hardware/ software requirements:

- At least 4 GB of RAM

- At least 1 GB of free disk space

- Java JDK 8+

All the examples and solutions to the exercises in this chapter can be found in this book's GitHub repository at `https://github.com/PacktPublishing/Continuous-Delivery-With-Docker-and-Jenkins-3rd-Edition/tree/main/Chapter06`.

Code in Action videos for this chapter can be viewed at `https://bit.ly/3rcffcz`.

Server clustering

So far, we have interacted with each of the machines individually. What we did was connect to the `localhost` Docker daemon server. We could have used the `-H` option in the `docker run` command to specify the address of the remote Docker, but that would still mean deploying our application to a single Docker host machine. In real life, however, if servers share the same physical location, we are not interested in which particular machine the service is deployed in. All we need is to have it accessible and replicated in many instances to support high availability. *How can we configure a set of machines to work that way?* This is the role of clustering.

In the following subsections, you will be introduced to the concept of server clustering and the Kubernetes environment, which is an example of cluster management software.

Introducing server clustering

A server cluster is a set of connected computers that work together in such a way that they can be used similarly to a single system. Servers are usually connected through the local network by a connection that's fast enough to ensure that the services that are being run are distributed. A simple server cluster is presented in the following diagram:

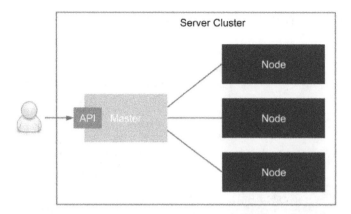

Figure 6.1 – Server clustering

A user accesses the cluster through a master host, which exposes the cluster API. There are multiple nodes that act as computing resources, which means that they are responsible for running applications. The master, on the other hand, is responsible for all other activities, such as the orchestration process, service discovery, load balancing, and node failure detection.

Introducing Kubernetes

Kubernetes is an open source cluster management system that was originally designed by Google. Looking at the popularity charts, it is a clear winner among other competitors, such as Docker Swarm and Apache Mesos. Its popularity has grown so fast that most cloud platforms provide Kubernetes out of the box. It's not Docker-native, but there are a lot of additional tools and integrations to make it work smoothly with the whole Docker ecosystem; for example, **kompose** can translate Docker Compose files into Kubernetes configurations.

> **Information**
>
> In the first edition of this book, I recommended Docker Compose and Docker Swarm for application dependency resolution and server clustering. While they're both good tools, Kubernetes' popularity has grown so high recently that I decided to use Kubernetes as the recommended approach and keep Docker-native tooling as an alternative.

Let's take a look at the simplified architecture of Kubernetes:

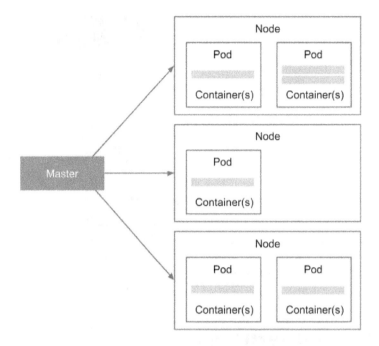

Figure 6.2 – Simplified Kubernetes architecture

The Kubernetes **control plane** (master), which is actually a set of cluster services, is responsible for enforcing the desired state of your applications. In other words, you specify your deployment setup in a declarative manner (four replicas of a web service exposing port 8080) and the control plane is responsible for making it happen. A Kubernetes Node, on the other hand, is a worker. You may see it just as a (Docker) container host with a special Kubernetes process (called kubelet) installed.

From the user's perspective, you provide a declarative deployment configuration in the form of a YAML file and pass it to the Kubernetes control plane through its API. Then, the control plane reads the configuration and installs the deployment. Kubernetes introduces the concept of a **Pod**, which represents a single deployment unit. The Pod contains Docker **containers**, which are scheduled together. While you can put multiple containers into a single Pod, in real-life scenarios, you will see that most Pods contain just a single Docker container. Pods are dynamically built and removed depending on the requirement changes that are expressed in the YAML configuration updates.

You will gain more practical knowledge about Kubernetes in later sections of this chapter, but first, let's name the features that make Kubernetes such a great environment.

Kubernetes features overview

Kubernetes provides a number of interesting features. Let's walk through the most important ones:

- **Container balancing**: Kubernetes takes care of the load balancing of Pods on nodes; you specify the number of replicas of your application and Kubernetes takes care of the rest.

- **Traffic load balancing**: When you have multiple replicas of your application, the Kubernetes service can load balance the traffic. In other words, you create a service with a single IP (or DNS) and Kubernetes takes care of load balancing the traffic to your application replicas.

- **Dynamic horizontal scaling**: Each deployment can be dynamically scaled up or down; you specify the number of application instances (or the rules for autoscaling) and Kubernetes starts/stops Pod replicas.

- **Failure recovery**: Pods (and nodes) are constantly monitored and if any of them fail, new Pods are started so that the declared number of replicas is constant.

- **Rolling updates**: An update to the configuration can be applied incrementally; for example, if we have 10 replicas and we would like to make a change, we can define a delay between the deployment to each replica. In such a case, when anything goes wrong, we never end up with a scenario where a replica isn't working correctly.

- **Storage orchestration**: Kubernetes can mount a storage system of your choice to your applications. Pods are stateless in nature and, therefore, Kubernetes integrates with a number of storage providers, such as Amazon **Elastic Block Storage** (**EBS**), **Google Compute Engine** (**GCE**) Persistent Disk, and Azure Data Storage.

- **Service discovery**: Kubernetes Pods are ephemeral in nature and their IPs are dynamically assigned, but Kubernetes provides DNS-based service discovery for this.

- **Run everywhere**: Kubernetes is an open source tool, and you have a lot of options of how to run it: on-premises, cloud infrastructure, or hybrid.

Now that we have some background about Kubernetes, let's see what it all looks like in practice, starting with the installation process.

Kubernetes installation

Kubernetes, just like Docker, consists of two parts: the client and the server. The client is a command-line tool named kubectl and it connects to the server part using the Kubernetes API. The server is much more complex and is as we described in the previous section. Obviously, to do anything with Kubernetes, you need both parts, so let's describe them one by one, starting with the client.

Kubernetes client

The Kubernetes client, kubectl, is a command-line application that allows you to perform operations on the Kubernetes cluster. The installation process depends on your operating system. You can check out the details on the official Kubernetes website: https://kubernetes.io/docs/tasks/tools/.

After you have successfully installed kubectl, you should be able to execute the following command:

```
$ kubectl version --client
Client Version: version.Info{Major:"1", Minor:"22",
GitVersion:"v1.22.4", ...
```

Now that you have the Kubernetes client configured, we can move on to the server.

Kubernetes server

There are multiple ways to set up a Kubernetes server. Which one you should use depends on your needs, but if you are completely new to Kubernetes, then I recommend starting from a local environment.

Local environment

Even though Kubernetes itself is a complex clustering system, there are a few tools that can simplify your local development setup. Let's walk through the options you have, which include Docker Desktop, kind, and minikube.

Docker Desktop

Docker Desktop is an application that is used to set up a local Docker environment on macOS or Windows. As you may remember from the previous chapters, the Docker daemon can only run natively on Linux, so for other operating systems, you need to get it running on a VM. Docker Desktop provides a super-intuitive way to do this, and luckily, it also supports the creation of Kubernetes clusters.

If you have Docker Desktop installed, then all you need to do is check the **Enable Kubernetes** box in the user interface, as shown in the following screenshot. From here, the Kubernetes cluster will start and `kubectl` will be configured:

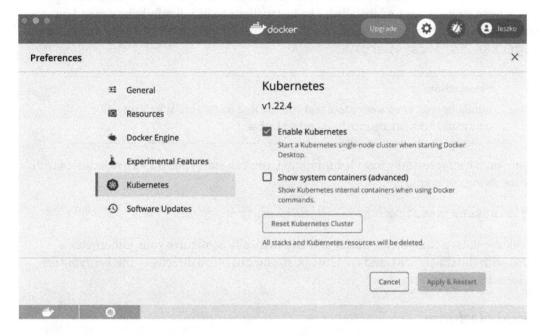

Figure 6.3 – Kubernetes in Docker Desktop

Starting from this point, you are ready to use the Kubernetes cluster.

kind

If you use the Linux operating system and can't, or just don't want to, use Docker Desktop, then your second simplest option is **kind** (short for **Kubernetes in Docker**). It's a tool for which the only requirement is to have Docker installed and configured.

After installing kind, you can start and configure your local Kubernetes cluster with this one command:

```
$ kind create cluster
```

> **Information**
>
> You can check the kind installation steps at `https://kind.sigs.k8s.io/docs/user/quick-start/`.

minikube

minikube is a command-line tool that starts a fully functional Kubernetes environment inside a VM. It is backed up by a VM hypervisor, so you need to have VirtualBox, Hyper-V, VMware, or a similar tool installed. The instructions to install minikube depend on your operating system, and you can find instructions for each at `https://minikube.sigs.k8s.io/docs/start/`.

> **Information**
>
> minikube is an open source tool that you can find on GitHub at `https://github.com/kubernetes/minikube`.

After you have successfully installed minikube, you can start your Kubernetes cluster with the following command:

```
$ minikube start
```

minikube starts a Kubernetes cluster and automatically configures your Kubernetes client with the cluster URL and credentials, so you can move directly to the *Verifying the Kubernetes setup* section.

Cloud platforms

Kubernetes has become so popular that most cloud computing platforms provide it as a service. The leader here is **Google Cloud Platform** (**GCP**), which allows you to create a Kubernetes cluster within a few minutes. Other cloud platforms, such as Microsoft Azure, **Amazon Web Services** (**AWS**), and IBM Cloud, also have Kubernetes in their portfolios. Let's take a closer look at the three most popular solutions—**GCP**, **Azure**, and **AWS**.

Google Cloud Platform

You can access GCP at `https://cloud.google.com/`. After creating an account, you should be able to open their web console (`https://console.cloud.google.com`). One of the services in their portfolio is called **Google Kubernetes Engine** (**GKE**).

You can create a Kubernetes cluster by clicking in the user interface or by using the GCP command-line tool, called `gcloud`.

> **Information**
>
> You can read how to install `gcloud` on your operating system at the official GCP website: `https://cloud.google.com/sdk/docs/install`.

To create a Kubernetes cluster using the command-line tool, it's enough to execute the following command:

```
$ gcloud container clusters create test-cluster
```

Apart from creating a Kubernetes cluster, it automatically configures kubectl.

Microsoft Azure

Microsoft Azure also offers a very quick Kubernetes setup thanks to **Azure Kubernetes Service (AKS)**. Like GCP, you can use either a web interface or a command-line tool to create a cluster.

> **Information**
>
> You can access the Azure web console at https://portal.azure.com/. To install the Azure command-line tool, check the installation guide on their official page at https://docs.microsoft.com/en-us/cli/azure/install-azure-cli.

To create a Kubernetes cluster using the Azure command-line tool, assuming you already have an Azure resource group created, it's enough to run the following command:

```
$ az aks create -n test-cluster -g test-resource-group
```

After a few seconds, your Kubernetes cluster should be ready. To configure kubectl, run the following command:

```
$ az aks get-credentials -n test-cluster -g test-resource-group
```

By doing this, you will have successfully set up a Kubernetes cluster and configured kubectl.

Amazon Web Services

AWS provides a managed Kubernetes service called Amazon **Elastic Kubernetes Service (EKS)**. You can start using it by accessing the AWS web console at https://console.aws.amazon.com/eks or using the AWS command-line tool.

> **Information**
>
> You can check all the information (and the installation guide) for the AWS command-line tool at its official website: https://docs.aws.amazon.com/cli/.

As you can see, using Kubernetes in the cloud is a relatively simple option. Sometimes, however, you may need to install an on-premises Kubernetes environment from scratch on your own server machines. Let's discuss this in the next section.

On-premises

Installing Kubernetes from scratch on your own servers makes sense if you don't want to depend on cloud platforms or if your corporate security policies don't allow it. The installation process is relatively complex and out of the scope of this book, but you can find all the details in the official documentation at `https://kubernetes.io/docs/setup/production-environment/`.

Now that we have the Kubernetes environment configured, we can check that `kubectl` is connected to the cluster correctly and that we are ready to start deploying our applications.

Verifying the Kubernetes setup

No matter which Kubernetes server installation you choose, you should already have everything configured and the Kubernetes client should be filled with the cluster's URL and credentials. You can check this with the following command:

```
$ kubectl cluster-info
Kubernetes control plane is running at https://kubernetes.
docker.internal:6443
CoreDNS is running at https://kubernetes.docker.internal:6443/
api/v1/namespaces/kube-system/services/kube-dns:dns/proxy
```

This is the output for the Docker Desktop scenario and is why you can see `localhost`. Your output may be slightly different and may include more entries. If you see no errors, then everything is correct, and we can start using Kubernetes to run applications.

Using Kubernetes

We have the whole Kubernetes environment ready and `kubectl` configured. This means that it's high time to finally present the power of Kubernetes and deploy our first application. Let's use the `leszko/calculator` Docker image that we built in the previous chapters and start it in multiple replicas on Kubernetes.

Deploying an application

In order to start a Docker container on Kubernetes, we need to prepare a deployment configuration as a YAML file. Let's name it deployment.yaml:

```
apiVersion: apps/v1
kind: Deployment                       (1)
metadata:
  name: calculator-deployment          (2)
  labels:
    app: calculator
spec:
  replicas: 3                          (3)
  selector:                            (4)
    matchLabels:
      app: calculator
  template:                            (5)
    metadata:
      labels:                          (6)
        app: calculator
    spec:
      containers:
      - name: calculator               (7)
        image: leszko/calculator       (8)
        ports:                         (9)
        - containerPort: 8080
```

In this YAML configuration, we have to ensure the following:

1. We have defined a Kubernetes resource of the Deployment type from the apps/v1 Kubernetes API version.

2. The unique deployment name is calculator-deployment.

3. We have defined that there should be exactly 3 of the same Pods created.

4. selector defines how Deployment finds Pods to manage, in this case, just by the label.

5. template defines the specification for each created Pod.

6. Each Pod is labeled with app: calculator.

7. Each Pod contains a Docker container named `calculator`.

8. A Docker container was created from the image called `leszko/calculator`.

9. The Pod exposes container port `8080`.

To install the deployment, run the following command:

```
$ kubectl apply -f deployment.yaml
```

You can check that the three Pods, each containing one Docker container, have been created:

```
$ kubectl get pods
NAME                                      READY STATUS    RESTARTS
AGE
calculator-deployment-dccdf8756-h216c 1/1     Running 0
1m
calculator-deployment-dccdf8756-tgw48 1/1     Running 0
1m
calculator-deployment-dccdf8756-vtwjz 1/1     Running 0
1m
```

Each Pod runs a Docker container. We can check its logs by using the following command:

```
$ kubectl logs pods/calculator-deployment-dccdf8756-h216c
```

You should see the familiar Spring logo and the logs of our Calculator web service.

Information

To look at an overview of `kubectl` commands, please check out the official guide: `https://kubernetes.io/docs/reference/kubectl/overview/`.

We have just performed our first deployment to Kubernetes, and with just a few lines of code, we have three replicas of our Calculator web service application. Now, let's see how we can use the application we deployed. For this, we'll need to understand the concept of a Kubernetes Service.

Deploying a Kubernetes Service

Each Pod has an IP address in the internal Kubernetes network, which means that you can already access each Calculator instance from another Pod running in the same Kubernetes cluster. But *how do we access our application from the outside?* That is the role of a Kubernetes Service.

The idea of Pods and Services is that Pods are mortal—they get terminated, and then they get restarted. The Kubernetes orchestrator only cares about the right number of Pod replicas, not about the Pod's identity. That's why, even though each Pod has an (internal) IP address, we should not stick to it or use it. Services, on the other hand, act as a frontend for Pods. They have IP addresses (and DNS names) that can be used. Let's look at the following diagram, which presents the idea of a Pod and Service:

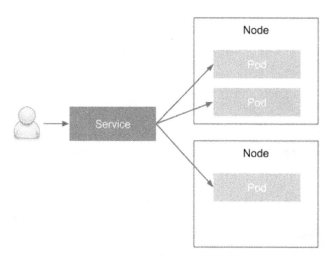

Figure 6.4 – Kubernetes Pod and Service

Pods are physically placed on different nodes, but you don't have to worry about this since Kubernetes takes care of the right orchestration and introduces the abstraction of a Pod and Service. The user accesses the Service, which load balances the traffic between the Pod replicas. Let's look at an example of how to create a service for our Calculator application.

Just like we did for the Deployment, we start from a YAML configuration file. Let's name it `service.yaml`:

```
apiVersion: v1
kind: Service
metadata:
  name: calculator-service
spec:
  type: NodePort
  selector:
    app: calculator
  ports:
  - port: 8080
```

This is a configuration for a simple service that load balances the traffic to all the Pods that meet the criteria we mentioned in `selector`. To install the service, run the following command:

```
$ kubectl apply -f service.yaml
```

You can then check that the service was correctly deployed by running the following command:

```
$ kubectl get service calculator-service
NAME                 TYPE      CLUSTER-IP      EXTERNAL-IP PORT(S)
AGE
calculator-service NodePort 10.19.248.154 <none>
8080:32259/TCP 13m
```

To check that the service points to the three Pod replicas we created in the previous section, run the following command:

```
$ kubectl describe service calculator-service | grep Endpoints
Endpoints: 10.16.1.5:8080,10.16.2.6:8080,10.16.2.7:8080
```

From the last two commands we ran, we can see that the service is available under the IP address of 10.19.248.154 and that it load balances the traffic to three Pods with the IPs of 10.16.1.5, 10.16.2.6, and 10.16.2.7. All of these IP addresses, for both the Service and Pod, are internal to the Kubernetes cluster network.

> **Information**
>
> To read more about Kubernetes Services, please visit the official Kubernetes website at https://kubernetes.io/docs/concepts/services-networking/service/.

In the next section, we'll take a look at how to access a service from outside the Kubernetes cluster.

Exposing an application

To understand how your application can be accessed from the outside, we need to start with the types of Kubernetes Services. You can use four different service types, as follows:

- **ClusterIP (default)**: The service has an internal IP only.

- **NodePort**: Exposes the service on the same port of each cluster node. In other words, each physical machine (which is a Kubernetes node) opens a port that is forwarded to the service. Then, you can access it by using `<NODE-IP>:<NODE-PORT>`.

- **LoadBalancer**: Creates an external load balancer and assigns a separate external IP for the service. Your Kubernetes cluster must support external load balancers, which works fine in the case of cloud platforms, but may not work if you use minikube.

- **ExternalName**: Exposes the service using a DNS name (specified by `externalName` in the spec).

If you use a Kubernetes instance that's been deployed on a cloud platform (for example, GKE), then the simplest way to expose your service is to use **LoadBalancer**. By doing this, GCP automatically assigns an external public IP for your service, which you can check with the `kubectl get service` command. If we had used it in our configuration, then you could have accessed the Calculator service at `http://<EXTERNAL-IP>:8080`.

While LoadBalancer seems to be the simplest solution, it has two drawbacks:

- First, it's not always available, for example, if you deployed on-premises Kubernetes or used minikube.

- Second, external public IPs are usually expensive. A different solution is to use a `NodePort` service, as we did in the previous section.

Now, let's see how we can access our service.

We can repeat the same command we ran already:

```
$ kubectl get service calculator-service
NAME                  TYPE       CLUSTER-IP     EXTERNAL-IP  PORT(S)
AGE
calculator-service NodePort 10.19.248.154 <none>
8080:32259/TCP 13m
```

You can see that port `32259` was selected as a node port. This means that we can access our Calculator service using that port and the IP of any of the Kubernetes nodes.

The IP address of your Kubernetes node depends on your installation. If you used Docker Desktop, then your node IP is `localhost`. In the case of minikube, you can check it with the `minikube ip` command. In the case of cloud platforms or the on-premises installation, you can check the IP addresses with the following command:

```
$ kubectl get nodes -o jsonpath='{ $.items[*].status.
addresses[?(@.type=="ExternalIP")].address }'
35.192.180.252 35.232.125.195 104.198.131.248
```

To check that you can access Calculator from the outside, run the following command:

```
$ curl <NODE-IP>:32047/sum?a=1\&b=2
3
```

We made an HTTP request to one of our Calculator container instances and it returned the right response, which means that we successfully deployed the application on Kubernetes.

> **Tip**
>
> The `kubectl` command offers a shortcut to create a service without using YAML. Instead of the configuration we used, you could just execute the following command:
>
> ```
> $ kubectl expose deployment calculator-deployment
> --type=NodePort --name=calculator-service.
> ```

What we've just learned gives us the necessary basics about Kubernetes. We can now use it for the staging and production environments and, therefore, include it in the continuous delivery process. Before we do so, however, let's look at a few more Kubernetes features that make it a great and useful tool.

Advanced Kubernetes

Kubernetes provides a way to dynamically modify your deployment during runtime. This is especially important if your application is already running in production and you need to support zero-downtime deployments. First, let's look at how to scale up an application and then present the general approach Kubernetes takes on any deployment changes.

Scaling an application

Let's imagine that our Calculator application is getting popular. People have started using it and the traffic is so high that the three Pod replicas are overloaded. *What can we do now?*

Luckily, `kubectl` provides a simple way to scale up and down deployments using the `scale` keyword. Let's scale our Calculator deployment to 5 instances:

```
$ kubectl scale --replicas 5 deployment calculator-deployment
```

That's it, our application is now scaled up:

```
$ kubectl get pods
NAME                                        READY STATUS    RESTARTS
AGE
calculator-deployment-dccdf8756-h216c 1/1    Running 0
19h
calculator-deployment-dccdf8756-j87kg 1/1    Running 0
36s
calculator-deployment-dccdf8756-tgw48 1/1    Running 0
19h
calculator-deployment-dccdf8756-vtwjz 1/1    Running 0
19h
calculator-deployment-dccdf8756-zw748 1/1    Running 0
36s
```

Note that, from now on, the service we created load balances the traffic to all 5 Calculator Pods. Also, note that you don't even need to wonder about which physical machine each Pod runs on, since this is covered by the Kubernetes orchestrator. All you have to think about is your desired number of application instances.

> **Information**
>
> Kubernetes also provides a way to autoscale your Pods, depending on its metrics. This feature is called the **HorizontalPodAutoscaler**, and you can read more about it at `https://kubernetes.io/docs/tasks/run-application/horizontal-pod-autoscale/`.

We have just seen how we can scale applications. Now, let's take a more generic look at how to update any part of a Kubernetes deployment.

Updating an application

Kubernetes takes care of updating your deployments. Let's make a change to `deployment.yaml` and add a new label to the Pod template:

```yaml
apiVersion: apps/v1
kind: Deployment
metadata:
  name: calculator-deployment
  labels:
    app: calculator
spec:
  replicas: 5
  selector:
    matchLabels:
      app: calculator
  template:
    metadata:
      labels:
        app: calculator
        label: label
    spec:
      containers:
      - name: calculator
        image: leszko/calculator
        ports:
        - containerPort: 8080
```

Now, if we repeat this and apply the same deployment, we can observe what happens with the Pods:

```
$ kubectl apply -f deployment.yaml
$ kubectl get pods
NAME                                         READY  STATUS
RESTARTS AGE
pod/calculator-deployment-7cc54cfc58-5rs9g 1/1     Running    0
7s
pod/calculator-deployment-7cc54cfc58-jcqlx 1/1     Running    0
4s
```

```
pod/calculator-deployment-7cc54cfc58-1sh7z  1/1   Running      0
4s
pod/calculator-deployment-7cc54cfc58-njbbc  1/1   Running      0
7s
pod/calculator-deployment-7cc54cfc58-pbthv  1/1   Running      0
7s
pod/calculator-deployment-dccdf8756-h216c   0/1   Terminating 0
20h
pod/calculator-deployment-dccdf8756-j87kg   0/1   Terminating 0
18m
pod/calculator-deployment-dccdf8756-tgw48   0/1   Terminating 0
20h
pod/calculator-deployment-dccdf8756-vtwjz   0/1   Terminating 0
20h
pod/calculator-deployment-dccdf8756-zw748   0/1   Terminating 0
18m
```

We can see that Kubernetes terminated all the old Pods and started the new ones.

Information

In our example, we modified the deployment of the YAML configuration, not the application itself. However, modifying the application is actually the same. If we make any change to the source code of the application, we need to build a new Docker image with the new version and then update this version in `deployment.yaml`.

Every time you change something and run `kubectl apply`, Kubernetes checks whether there is any change between the existing state and the YAML configuration, and then, if needed, it performs the update operation we described previously.

This is all well and good, but if Kubernetes suddenly terminates all Pods, we may end up in a situation where all the old Pods are already killed and none of the new Pods are ready yet. This would make our application unavailable for a moment. *How do we ensure zero-downtime deployments?* That's the role of rolling updates.

Rolling updates

A rolling update entails incrementally terminating old instances and starting new ones. In other words, the workflow is as follows:

1. Terminate one of the old Pods.

2. Start a new Pod.

3. Wait until the new Pod is ready.

4. Repeat *step 1* until all old instances are replaced.

> **Information**
>
> The concept of a rolling update works correctly only if the new application version is backward compatible with the old application version. Otherwise, we risk having two different incompatible versions at the same time.

To configure it, we need to add the `RollingUpdate` strategy to our deployment and specify `readinessProbe`, which makes Kubernetes aware when the Pod is ready. Let's modify `deployment.yaml`:

```yaml
apiVersion: apps/v1
kind: Deployment
metadata:
  name: calculator-deployment
  labels:
    app: calculator
spec:
  replicas: 5
  strategy:
    type: RollingUpdate
    rollingUpdate:
      maxUnavailable: 25%
      maxSurge: 0
  selector:
    matchLabels:
      app: calculator
  template:
    metadata:
      labels:
```

```
      app: calculator
  spec:
    containers:
    - name: calculator
      image: leszko/calculator
      ports:
      - containerPort: 8080
      readinessProbe:
        httpGet:
          path: /sum?a=1&b=2
          port: 8080
```

Let's explain the parameters we used in our configuration:

- `maxUnavailable`: The maximum number of Pods that can be unavailable during the update process; in our case, Kubernetes won't terminate at the same time when there's more than one Pod (*75% * 5* desired replicas).

- `maxSurge`: The maximum number of Pods that can be created over the desired number of Pods; in our case, Kubernetes won't create any new Pods before terminating an old one.

- `path` and `port`: The endpoint of the container to check for readiness; an HTTP GET request is sent to `<POD-IP>:8080/sum?a=1&b=2` and when it finally returns `200` as the HTTP status code, the Pod is marked as *ready*.

> **Tip**
>
> By modifying the `maxUnavailable` and `maxSurge` parameters, we can decide whether Kubernetes first starts new Pods and later terminates old ones or, as we did in our case, first terminates old Pods and later starts new ones.

We can now apply the deployment and observe that the Pods are updated one by one:

```
$ kubectl apply -f deployment.yaml
$ kubectl get pods
NAME                                         READY  STATUS
RESTARTS AGE
calculator-deployment-78fd7b57b8-npphx 0/1    Running    0
4s
calculator-deployment-7cc54cfc58-5rs9g 1/1    Running    0
3h
```

```
calculator-deployment-7cc54cfc58-jcqlx 0/1    Terminating 0
3h
calculator-deployment-7cc54cfc58-lsh7z 1/1    Running     0
3h
calculator-deployment-7cc54cfc58-njbbc 1/1    Running     0
3h
calculator-deployment-7cc54cfc58-pbthv 1/1    Running     0
3h
```

That's it, we have just configured a rolling update for our Calculator deployment, which means that we can provide zero-downtime releases.

> **Information**
>
> Kubernetes also provides a different way of running applications. You can use `StatefulSet` instead of `Deployment`, and then the rolling update is always enabled (even without specifying any additional strategy).

Rolling updates are especially important in the context of continuous delivery, because if we deploy very often, then we definitely can't afford any downtime.

> **Tip**
>
> After playing with Kubernetes, it's good to perform the cleanup to remove all the resources we created. In our case, we can execute the following commands to remove the service and deployment we created:
>
> ```
> $ kubectl delete -f service.yaml
> $ kubectl delete -f deployment.yaml
> ```

We've already presented all the Kubernetes features that are needed for the continuous delivery process. Let's look at a short summary and add a few words about other useful features.

Kubernetes objects and workloads

The execution unit in Kubernetes is always a Pod, which contains one or more (Docker) containers. There are multiple different resource types to orchestrate Pods:

- **Deployment**: This is the most common workload, which manages the life cycle of the desired number of replicated Pods.

- **StatefulSet**: This is a specialized Pod controller that guarantees the ordering and uniqueness of Pods. It is usually associated with data-oriented applications (in which it's not enough to say, *my desired number of replicas is 3*, as in the case of a Deployment, but rather, *I want exactly 3 replicas, with always the same predictable Pod names, and always started in the same order*).

- **DaemonSet**: This is a specialized Pod controller that runs a copy of a Pod on each Kubernetes node.

- **Job/CronJob**: This is a workflow that's dedicated to task-based operations in which containers are expected to exist successfully.

> **Information**
>
> You may also find a Kubernetes resource called **ReplicationController**, which is deprecated and has been replaced by Deployment.

Apart from Pod management, there are other Kubernetes objects. The most useful ones that you may often encounter are as follows:

- **Service**: A component that acts as an internal load balancer for Pods.

- **ConfigMap**: This decouples configuration from the image content; it can be any data that's defined separately from the image and then mounted onto the container's filesystem.

- **Secret**: This allows you to store sensitive information, such as passwords.

- **PersistentVolume/PersistentVolumeClaim**: These allow you to mount a persistent volume into a (stateless) container's filesystem.

Actually, there are many more objects available, and you can even create your own resource definitions. However, the ones we've mentioned here are the most frequently used in practice.

We already have a good understanding of clustering in Kubernetes, but Kubernetes isn't just about workloads and scaling. It can also help with resolving dependencies between applications. In the next section, we will approach this topic and describe application dependencies in the context of Kubernetes and the continuous delivery process.

Application dependencies

Life is easy without dependencies. In real life, however, almost every application links to a database, cache, messaging system, or another application. In the case of (micro) service architecture, each service needs a bunch of other services to do its work. The monolithic architecture does not eliminate the issue—an application usually has some dependencies, at least to the database.

Imagine a newcomer joining your development team; *how much time does it take to set up the entire development environment and run the application with all its dependencies?*

When it comes to automated acceptance testing, the dependencies issue is no longer only a matter of convenience—it becomes a necessity. While, during unit testing, we could mock the dependencies, the acceptance testing suite requires a complete environment. *How do we set it up quickly and in a repeatable manner?* Luckily, Kubernetes can help thanks to its built-in DNS resolution for Services and Pods.

The Kubernetes DNS resolution

Let's present the Kubernetes DNS resolution with a real-life scenario. Let's say we would like to deploy a caching service as a separate application and make it available for other services. One of the best in-memory caching solutions is Hazelcast, so let's use it here. In the case of the Calculator application, we need `Deployment` and `Service`. Let's define them both in one file, `hazelcast.yaml`:

```yaml
apiVersion: apps/v1
kind: Deployment
metadata:
  name: hazelcast
  labels:
    app: hazelcast
spec:
  replicas: 1
  selector:
    matchLabels:
      app: hazelcast
  template:
    metadata:
      labels:
        app: hazelcast
    spec:
```

```
        containers:
        - name: hazelcast
          image: hazelcast/hazelcast:5.0.2
          ports:
          - containerPort: 5701
---
apiVersion: v1
kind: Service
metadata:
  name: hazelcast
spec:
  selector:
    app: hazelcast
  ports:
  - port: 5701
```

Similar to what we did previously for the Calculator application, we will now define the Hazelcast configuration. Let's start it in the same way:

```
$ kubectl apply -f hazelcast.yaml
```

After a few seconds, the Hazelcast caching application should start. You can check its Pod logs with the `kubectl logs` command. We also created a service of a default type (`ClusterIP`, which is only exposed inside the same Kubernetes cluster).

So far, so good—we did nothing different from what we've already seen in the case of the Calculator application. Now comes the most interesting part. Kubernetes provides a way of resolving a service IP using the service name. What's even more interesting is that we know the `Service` name upfront—in our case, it's always `hazelcast`. So, if we use this as the cache address in our application, the dependency will be automatically resolved.

> **Information**
> Actually, Kubernetes DNS resolution is even more powerful, and it can resolve Services in a different Kubernetes namespace. Read more at `https://kubernetes.io/docs/concepts/services-networking/dns-pod-service/`.

Before we show you how to implement caching inside the Calculator application, let's take a moment to overview the system we will build.

Multiapplication system overview

We already have the Hazelcast server deployed on Kubernetes. Before we modify our Calculator application so that we can use it as a caching provider, let's take a look at a diagram of the complete system we want to build:

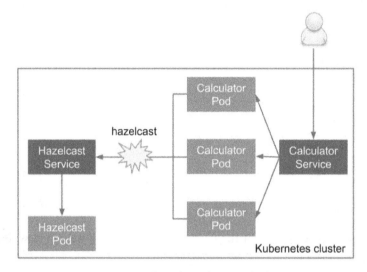

Figure 6.5 – Sample multiapplication deployment

The user uses the **Calculator Service**, which load balances the traffic to a **Calculator Pod**. Then, the **Calculator Pod** connects to the **Hazelcast Service** (using its name, hazelcast). The **Hazelcast Service** redirects to the **Hazelcast Pod**.

If you look at the diagram, you can see that we have just deployed the Hazelcast part (**Hazelcast Service** and **Hazelcast Pod**). We also deployed the Calculator part (**Calculator Service** and **Calculator Pod**) in the previous section. The final missing part is the Calculator code to use Hazelcast. Let's implement it now.

Multiapplication system implementation

To implement caching with Hazelcast in our Calculator application, we need to do the following:

1. Add the Hazelcast client library to Gradle.
2. Add the Hazelcast cache configuration.
3. Add Spring Boot caching.
4. Build a Docker image.

Let's proceed step by step.

Adding the Hazelcast client library to Gradle

In the `build.gradle` file, add the following configuration to the `dependencies` section:

```
implementation 'com.hazelcast:hazelcast:5.0.2'
```

This adds the Java libraries that take care of communication with the Hazelcast server.

Adding the Hazelcast cache configuration

Add the following parts to the `src/main/java/com/leszko/calculator/CalculatorApplication.java` file:

```java
package com.leszko.calculator;
import com.hazelcast.client.config.ClientConfig;
import org.springframework.boot.SpringApplication;
import org.springframework.boot.autoconfigure.SpringBootApplication;
import org.springframework.cache.annotation.EnableCaching;
import org.springframework.context.annotation.Bean;


@SpringBootApplication
@EnableCaching
public class CalculatorApplication {

    public static void main(String[] args) {
        SpringApplication.run(CalculatorApplication.class, args);
    }


    @Bean
    public ClientConfig hazelcastClientConfig() {
        ClientConfig clientConfig = new ClientConfig();
        clientConfig.getNetworkConfig().addAddress("hazelcast");
        return clientConfig;
    }
}
```

This is a standard Spring cache configuration. Note that for the Hazelcast server address, we use `hazelcast`, which is automatically available thanks to the Kubernetes DNS resolution.

> **Tip**
>
> In real life, if you use Hazelcast, you don't even need to specify the service name, since Hazelcast provides an autodiscovery plugin dedicated to the Kubernetes environment. Read more at `https://docs.hazelcast.com/hazelcast/latest/deploy/deploying-in-kubernetes.html`.

We also need to remove the Spring context test automatically created by Spring Initializr, `src/test/java/com/leszko/calculator/CalculatorApplicationTests.java`.

Next, let's add caching to the Spring Boot service.

Adding Spring Boot caching

Now that the cache is configured, we can finally add caching to our web service. In order to do this, we need to change the `src/main/java/com/leszko/calculator/Calculator.java` file so that it looks as follows:

```java
package com.leszko.calculator;

import org.springframework.cache.annotation.Cacheable;
import org.springframework.stereotype.Service;

@Service
public class Calculator {
    @Cacheable("sum")
    public int sum(int a, int b) {
        try {
            Thread.sleep(3000);
        }
        catch (InterruptedException e) {
            e.printStackTrace();
        }
```

```
        return a + b;
    }
}
```

We added the `@Cacheable` annotation to make Spring automatically cache every call of the `sum()` method. We also added sleeping for 3 seconds, just for the purpose of testing, so that we could see that the cache works correctly.

From now on, the sum calculations are cached in Hazelcast, and when we call the `/sum` endpoint of the Calculator web service, it will first try to retrieve the result from the cache. Now, let's build our application.

Building a Docker image

As the next step, we need to remove the Spring default context test, `src/test/java/com/leszko/calculator/CalculatorApplicationTests.java` (to avoid failing because of the missing Hazelcast dependency).

Now, we can rebuild the Calculator application and the Docker image with a new tag. Then, we will push it to Docker Hub once more:

```
$ ./gradlew build
$ docker build -t leszko/calculator:caching .
$ docker push leszko/calculator:caching
```

Obviously, you should change `leszko` to your Docker Hub account.

The application is ready, so let's test it all together on Kubernetes.

Multiapplication system testing

We should already have the Hazelcast caching server deployed on Kubernetes. Now, let's change the deployment for the Calculator application to use the `leszko/calculator:caching` Docker image. You need to modify `image` in the `deployment.yaml` file:

```
image: leszko/calculator:caching
```

Then, apply the Calculator deployment and service:

```
$ kubectl apply -f deployment.yaml
$ kubectl apply -f service.yaml
```

Let's repeat the `curl` operation we did before:

```
$ curl <NODE-IP>:<NODE-PORT>/sum?a=1\&b=2
```

The first time you execute it, it should reply in 3 seconds, but all subsequent calls should be instant, which means that caching works correctly.

> **Tip**
>
> If you're interested, you can also check the logs of the Calculator Pod. You should see some logs there that confirm that the application is connected to the Hazelcast server:
>
> ```
> Members [1] {
>     Member [10.16.2.15]:5701 - 3fca574b-bbdb-4c14-
> ac9d-73c45f56b300
> }
> ```

You can probably already see how we could perform acceptance testing on a multicontainer system. All we need is an acceptance test specification for the whole system. Then, we could deploy the complete system into the Kubernetes staging environment and run a suite of acceptance tests against it. We'll talk about this in more detail in *Chapter 8, Continuous Delivery Pipeline*.

> **Information**
>
> In our example, the dependent service was related to caching, which doesn't really change the functional acceptance tests we created in *Chapter 5, Automated Acceptance Testing*.

That's all we need to know about how to approach dependent applications that are deployed on the Kubernetes cluster in the context of continuous delivery. Nevertheless, before we close this chapter, let's write a few words about Kubernetes' competitors, that is, other popular cluster management systems.

Alternative cluster management systems

Kubernetes is not the only system that can be used to cluster Docker containers. Even though it's currently the most popular one, there may be some valid reasons to use different software. Let's walk through the alternatives.

Docker Swarm

Docker Swarm is a native clustering system for Docker that turns a set of Docker hosts into one consistent cluster, called a **swarm**. Each host connected to the swarm plays the role of a manager or a worker (there must be at least one manager in a cluster). Technically, the physical location of the machines does not matter; however, it's reasonable to have all Docker hosts inside one local network; otherwise, managing operations (or reaching a consensus between multiple managers) can take a significant amount of time.

> **Information**
>
> Since Docker 1.12, Docker Swarm is natively integrated into Docker Engine in swarm mode. In older versions, it was necessary to run the swarm container on each of the hosts to provide the clustering functionality.

Let's look at the following diagram, which presents the terminology and the Docker Swarm clustering process:

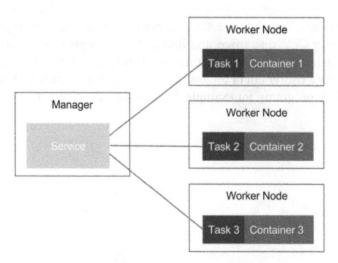

Figure 6.6 – Docker Swarm

In Docker Swarm mode, a running image is called a **Service**, as opposed to a **container**, which is run on a single Docker host. One service runs a specified number of **tasks**. A task is an atomic scheduling unit of the swarm that holds the information about the container and the command that should be run inside the container. A **replica** is each container that is run on the node. The number of replicas is the expected number of all containers for the given service.

We start by specifying a service, the Docker image, and the number of replicas. The manager automatically assigns tasks to worker nodes. Obviously, each replicated container is run from the same Docker image. In the context of the presented flow, Docker Swarm can be viewed as a layer on top of the Docker Engine mechanism that is responsible for container orchestration.

> **Information**
>
> In the first edition of this book, Docker Swarm was used for all the examples that were provided. So, if Docker Swarm is your clustering system of choice, you may want to read the first edition.

Another alternative to Kubernetes is Apache Mesos. Let's talk about it now.

Apache Mesos

Apache Mesos is an open source scheduling and clustering system that was started at the University of California, Berkeley, in 2009, long before Docker emerged. It provides an abstraction layer over CPU, disk space, and RAM. One of the great advantages of Mesos is that it supports any Linux application, but not necessarily (Docker) containers. This is why it's possible to create a cluster out of thousands of machines and use it for both Docker containers and other programs, for example, Hadoop-based calculations.

Let's look at the following diagram, which presents the Mesos architecture:

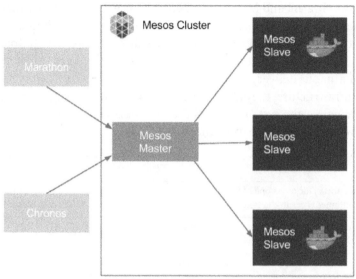

Figure 6.7 – Apache Mesos

Apache Mesos, similar to other clustering systems, has the master-slave architecture. It uses node agents that have been installed on every node for communication, and it provides two types of schedulers:

- **Chronos**: For cron-style repeating tasks
- **Marathon**: To provide a REST API to orchestrate services and containers

Apache Mesos is very mature compared to other clustering systems, and it has been adopted in a large number of organizations, such as Twitter, Uber, and CERN.

Comparing features

Kubernetes, Docker Swarm, and Mesos are all good choices for the cluster management system. All of them are free and open source, and all of them provide important cluster management features, such as load balancing, service discovery, distributed storage, failure recovery, monitoring, secret management, and rolling updates. All of them can also be used in the continuous delivery process without huge differences. This is because, in the Dockerized infrastructure, they all address the same issue—the clustering of Docker containers. Nevertheless, the systems are not exactly the same. Let's take a look at the following table, which presents the differences:

	Kubernetes	Docker Swarm	Apache Mesos
Docker support	Supports Docker as one of the container types in the Pod	Native	Mesos agents (slaves) can be configured to host Docker containers
Application types	Containerized applications (Docker, RKT, and others)	Docker images	Any application that can be run on Linux (also containers)
Application definition	Deployments, StatefulSets, and Services	Docker Compose configuration	Application groups formed in the tree structure
Setup process	Depending on the infrastructure, it may require running one command or many complex operations	Very simple	Fairly involved; it requires configuring Mesos, Marathon, Chronos, ZooKeeper, and Docker Support
API	REST API	Docker REST API	Chronos and Marathon REST API
User interface	Console tools, native web user interface (Kubernetes Dashboard)	Docker console client and third-party web applications, such as Shipyard	Official web interfaces for Mesos, Marathon, and Chronos
Cloud integration	Cloud-native support from most providers (Azure, AWS, GCP, and others)	Manual installation required	Support from most cloud providers
Maximum cluster size	1,000 nodes	1,000 nodes	50,000 nodes
Autoscaling	HorizontalPodAutoscaling based on the observed metrics	Not available	Marathon provides autoscaling based on resource (CPU/memory) consumption, number of requests per second, and queue length

Obviously, apart from Kubernetes, Docker Swarm, and Apache Mesos, there are other clustering systems available on the market. Especially in the era of cloud platforms, there are very popular platform-specific systems, for example, Amazon **Elastic Container Service (ECS)**. The good news is that if you understand the idea of clustering Docker containers, then using another system won't be difficult for you.

Summary

In this chapter, we took a look at the clustering methods for Docker environments that allow you to set up complete staging and production environments. Let's go over some of the key takeaways from this chapter:

- Clustering is a method of configuring a set of machines in a way that, in many respects, can be viewed as a single system.
- Kubernetes is the most popular clustering system for Docker.
- Kubernetes consists of the Kubernetes server and the Kubernetes client (`kubectl`).
- The Kubernetes server can be installed locally (through minikube or Docker Desktop), on the cloud platform (AKS, GKE, or EKS), or manually on a group of servers. Kubernetes uses YAML configurations to deploy applications.
- Kubernetes provides features such as scaling and rolling updates out of the box.
- Kubernetes provides DNS resolution, which can help when you're deploying systems that consist of multiple dependent applications.
- The most popular clustering systems that support Docker are Kubernetes, Docker Swarm, and Apache Mesos.

In the next chapter, we will describe the configuration management part of the continuous delivery pipeline.

Exercises

In this chapter, we have covered Kubernetes and the clustering process in detail. In order to enhance this knowledge, we recommend the following exercises:

1. Run a `hello world` application on the Kubernetes cluster:
    I. The `hello world` application can look exactly the same as the one we described in the exercises for *Chapter 2, Introducing Docker*.
    II. Deploy the application with three replicas.

    III. Expose the application with the `NodePort` service.

    IV. Make a request (using `curl`) to the application.

2. Implement a new feature, *Goodbye World!*, and deploy it using a rolling update:

    I.   This feature can be added as a new endpoint, `/bye`, which always returns *Goodbye World!*.

    II. Rebuild a Docker image with a new version tag.

    III. Use the `RollingUpdate` strategy and `readinessProbe`.

    IV. Observe the rolling update procedure.

    V.   Make a request (using `curl`) to the application.

Questions

To verify your knowledge from this chapter, please answer the following questions:

1. What is a server cluster?
2. What is the difference between a Kubernetes control plane and Kubernetes Node?
3. Name at least three cloud platforms that provide a Kubernetes environment out of the box.
4. What is the difference between a Kubernetes deployment and service?
5. What is the Kubernetes command for scaling deployments?
6. Name at least two cluster management systems other than Kubernetes.

Further reading

To find out more about Kubernetes, please refer to the following resources:

- **Kubernetes official documentation**: `https://kubernetes.io/docs/home/`
- **Nigel Poulton: The Kubernetes Book** (`https://leanpub.com/thekubernetesbook`)

Section 3 – Deploying an Application

In this section, we will cover how to release an application on a Docker production server using configuration management tools such as Chef and Ansible, as well as crucial parts of the continuous delivery process. We will also address more difficult real-life scenarios after building a complete pipeline.

The following chapters are covered in this section:

- *Chapter 7, Configuration Management with Ansible*
- *Chapter 8, Continuous Delivery Pipeline*
- *Chapter 9, Advanced Continuous Delivery*

7

Configuration Management with Ansible

We have already covered the two most crucial phases of the continuous delivery process: the commit phase and automated acceptance testing. We also explained how to cluster your environments for both your application and Jenkins agents. In this chapter, we will focus on configuration management, which connects the virtual containerized environment to the real server infrastructure.

This chapter will cover the following points:

- Introducing configuration management
- Installing Ansible
- Using Ansible
- Deployment with Ansible
- Ansible with Docker and Kubernetes
- Introducing infrastructure as code
- Introducing Terraform

Technical requirements

To follow along with the instructions in this chapter, you'll need the following hardware/software:

- Java 8+
- Python
- Remote machines with the Ubuntu operating system and SSH server installed
- An AWS account

All the examples and solutions to the exercises can be found on GitHub at `https://github.com/PacktPublishing/Continuous-Delivery-With-Docker-and-Jenkins-3rd-Edition/tree/main/Chapter07`.

Code in Action videos for this chapter can be viewed at `https://bit.ly/3JkcGLE`.

Introducing configuration management

Configuration management is the process of controlling configuration changes in such a way that the system maintains integrity over time. Even though the term did not originate in the IT industry, currently, it is broadly used to refer to software and hardware. In this context, it concerns the following aspects:

- **Application configuration**: This involves software properties that decide how the system works, which are usually expressed in the form of flags or properties files passed to the application, for example, the database address, the maximum chunk size for file processing, or the logging level. They can be applied during different development phases: build, package, deploy, or run.

- **Server configuration**: This defines what dependencies should be installed on each server and specifies the way applications are orchestrated (which application is run on which server, and in how many instances).

- **Infrastructure configuration**: This involves server infrastructure and environment configuration. If you use on-premises servers, then this part is related to the manual hardware and network installation; if you use cloud solutions, then this part can be automated with the **infrastructure as code** (**IaC**)e approach.

As an example, we can think of the calculator web service, which uses the Hazelcast server. Let's look at the following diagram, which presents how configuration management works:

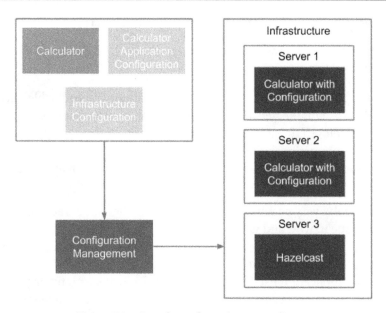

Figure 7.1 – Sample configuration management

The configuration management tool reads the configuration file and prepares the environment. It installs dependent tools and libraries and deploys the applications to multiple instances. Additionally, in the case of cloud deployment, it can provide the necessary infrastructure.

In the preceding example, **Infrastructure Configuration** specifies the required servers and S**erver Configuration** defines that the **Calculator** service should be deployed in two instances, on **Server 1** and **Server 2**, and that the Hazelcast service should be installed on **Server 3**. **Calculator Application Configuration** specifies the port and the address of the Hazelcast server so that the services can communicate.

> **Information**
> The configuration can differ, depending on the type of the environment (QA, staging, or production); for example, server addresses can be different.

There are many approaches to configuration management, but before we look into concrete solutions, let's comment on what characteristics a good configuration management tool should have.

Traits of good configuration management

What should a modern configuration management solution look like? Let's walk through the most important factors:

- **Automation**: Each environment should be automatically reproducible, including the operating system, the network configuration, the software installed, and the applications deployed. In such an approach, fixing production issues means nothing more than an automatic rebuild of the environment. What's more, it simplifies server replications and ensures that the staging and production environments are exactly the same.

- **Version control**: Every change in the configuration should be tracked, so that we know who made it, why, and when. Usually, that means keeping the configuration in the source code repository, either with the code or in a separate place. The former solution is recommended because configuration properties have a different life cycle than the application itself. Version control also helps with fixing production issues; the configuration can always be rolled back to the previous version, and the environment automatically rebuilt. The only exception to the version control-based solution is storing credentials and other sensitive information; these should never be checked in.

- **Incremental changes**: Applying a change in the configuration should not require rebuilding the whole environment. On the contrary, a small change in the configuration should only change the related part of the infrastructure.

- **Server provisioning**: Thanks to automation, adding a new server should be as quick as adding its address to the configuration (and executing one command).

- **Security**: The access to both the configuration management tool and the machines under its control should be well secured. When using the SSH protocol for communication, the access to the keys or credentials needs to be well protected.

- **Simplicity**: Every member of the team should be able to read the configuration, make a change, and apply it to the environment. The properties themselves should also be kept as simple as possible, and the ones that are not subject to change are better off kept hardcoded.

It is important to keep these points in mind while creating the configuration, and even beforehand while choosing the right configuration management tool.

Overview of configuration management tools

In the classic sense, before the cloud era, configuration management referred to the process that started when all the servers were already in place. So, the starting point was a set of IP addresses with machines accessible via SSH. For that purpose, the most popular configuration management tools are Ansible, Puppet, and Chef. Each of them is a good choice; they are all open source products with free basic versions and paid enterprise editions. The most important differences between them are as follows:

- **Configuration language**: Chef uses Ruby, Puppet uses its own DSL (based on Ruby), and Ansible uses YAML.

- **Agent-based**: Puppet and Chef use agents for communication, which means that each managed server needs to have a special tool installed. Ansible, on the other hand, is agentless and uses the standard SSH protocol for communication.

The agentless feature is a significant advantage because it implies no need to install anything on servers. What's more, Ansible is quickly trending upward, which is why it was chosen for this book. Nevertheless, other tools can also be used successfully for the continuous delivery process.

Together with cloud transformation, the meaning of configuration management widened and started to include what is called **IaC**. As the input, you no longer need a set of IP addresses, but it's enough to provide the credentials to your favorite cloud provider. Then, IaC tools can provision servers for you. What's more, each cloud provider offers a portfolio of services, so in many cases, you don't even need to provision bare-metal servers, but directly use cloud services. While you can still use Ansible, Puppet, or Chef for that purpose, there is a tool called Terraform that is dedicated to the IaC use case.

Let's first describe the classic approach to configuration management with Ansible, and then walk through the IaC solution using Terraform.

Installing Ansible

Ansible is an open source, agentless automation engine for software provisioning, configuration management, and application deployment. Its first release was in 2012, and its basic version is free for both personal and commercial use. The enterprise version is called **Ansible Tower**, which provides GUI management and dashboards, the REST API, role-based access control, and some more features.

We will present the installation process and a description of how Ansible can be used separately, as well as in conjunction with Docker.

Ansible server requirements

Ansible uses the SSH protocol for communication and has no special requirements regarding the machine it manages. There is also no central master server, so it's enough to install the Ansible client tool anywhere; we can then use it to manage the whole infrastructure.

> **Information**
>
> The only requirement for the machines being managed is to have the Python tool (and obviously, the SSH server) installed. These tools are, however, almost always available on any server by default.

Ansible installation

The installation instructions will differ depending on the operating system. In the case of Ubuntu, it's enough to run the following commands:

```
$ sudo apt-get install software-properties-common
$ sudo apt-add-repository ppa:ansible/ansible
$ sudo apt-get update
$ sudo apt-get install ansible
```

> **Information**
>
> You can find the installation guides for all the operating systems on the official Ansible page, at https://docs.ansible.com/ansible/latest/installation_guide/intro_installation.html.

After the installation process is complete, we can execute the `ansible` command to check that everything was installed successfully:

```
$ ansible --version
ansible [core 2.12.2]
  config file = /etc/ansible/ansible.cfg
...
```

Using Ansible

In order to use Ansible, we first need to define the inventory, which represents the available resources. Then, we will be able to either execute a single command or define a set of tasks using the Ansible playbook.

Creating an inventory

An inventory is a list of all the servers that are managed by Ansible. Each server requires nothing more than the Python interpreter and the SSH server installed. By default, Ansible assumes that SSH keys are used for authentication; however, it is also possible to use a username and password by adding the --ask-pass option to the Ansible commands.

> **Tip**
>
> SSH keys can be generated with the ssh-keygen tool, and they are usually stored in the ~/.ssh directory.

The inventory is defined by default in the /etc/ansible/hosts file (but its location can be defined with the -i parameter), and it has the following structure:

```
[group_name]
<server1_address>
<server2_address>
...
```

> **Tip**
>
> The inventory syntax also accepts ranges of servers, for example, www[01-22].company.com. The SSH port should also be specified if it's anything other than 22 (the default).

There can be many groups in the inventory file. As an example, let's define two machines in one group of servers:

```
[webservers]
192.168.64.12
192.168.64.13
```

We can also create the configuration with server aliases and specify the remote user:

```
[webservers]
web1 ansible_host=192.168.64.12 ansible_user=ubuntu
web2 ansible_host=192.168.64.13 ansible_user=ubuntu
```

The preceding file defines a group called `webservers`, which consists of two servers. The Ansible client will log into both of them as the user `ubuntu`. When we have the inventory created, let's discover how we can use it to execute the same command on many servers.

> **Information**
>
> Ansible offers the possibility to dynamically pull the inventory from a cloud provider (for example, Amazon EC2/Eucalyptus), LDAP, or Cobbler. Read more about dynamic inventories at `https://docs.ansible.com/ ansible/latest/user_guide/intro_dynamic_inventory. html`.

Ad hoc commands

The simplest command we can run is a ping on all servers. Assuming that we have two remote machines (`192.168.64.12` and `192.168.64.13`) with SSH servers configured and the inventory file (as defined in the last section), let's execute the `ping` command:

```
$ ansible all -m ping
web1 | SUCCESS => {
    "ansible_facts": {
        "discovered_interpreter_python": "/usr/bin/python3"
    },
    "changed": false,
    "ping": "pong"
}
web2 | SUCCESS => {
    "ansible_facts": {
        "discovered_interpreter_python": "/usr/bin/python3"
    },
    "changed": false,
    "ping": "pong"
}
```

We used the `-m <module_name>` option, which allows for specifying the module that should be executed on the remote hosts. The result is successful, which means that the servers are reachable, and the authentication is configured correctly.

Note that we used `all`, so that all servers would be addressed, but we could also call them by the `webservers` group name, or by the single host alias. As a second example, let's execute a shell command on only one of the servers:

```
$ ansible web1 -a "/bin/echo hello"
web1 | CHANGED | rc=0 >>
hello
```

The `-a <arguments>` option specifies the arguments that are passed to the Ansible module. In this case, we didn't specify the module, so the arguments are executed as a shell Unix command. The result was successful, and `hello` was printed.

> **Tip**
>
> If the `ansible` command is connecting to the server for the first time (or if the server is reinstalled), then we are prompted with the key confirmation message (the SSH message, when the host is not present in `known_hosts`). Since it may interrupt an automated script, we can disable the prompt message by uncommenting `host_key_checking = False` in the `/etc/ansible/ansible.cfg` file, or by setting the environment variable, `ANSIBLE_HOST_KEY_CHECKING=False`.

In its simplistic form, the Ansible ad hoc command syntax looks as follows:

```
$ ansible <target> -m <module_name> -a <module_arguments>
```

The purpose of ad hoc commands is to do something quickly when it is not necessary to repeat it. For example, we may want to check whether a server is alive or power off all the machines for the Christmas break. This mechanism can be seen as a command execution on a group of machines, with the additional syntax simplification provided by the modules. The real power of Ansible automation, however, lies in playbooks.

Playbooks

An **Ansible playbook** is a configuration file that describes how servers should be configured. It provides a way to define a sequence of tasks that should be performed on each of the machines. A playbook is expressed in the YAML configuration language, which makes it human-readable and easy to understand. Let's start with a sample playbook, and then see how we can use it.

Defining a playbook

A playbook is composed of one or many plays. Each play contains a host group name, tasks to perform, and configuration details (for example, the remote username or access rights). An example playbook might look like this:

```
---
- hosts: web1
  become: yes
  become_method: sudo
  tasks:
  - name: ensure apache is at the latest version
    apt: name=apache2 state=latest
  - name: ensure apache is running
    service: name=apache2 state=started enabled=yes
```

This configuration contains one play, which performs the following:

- Only executes on the web1 host
- Gains root access using the sudo command
- Executes two tasks:

  - **Installing the latest version of apache2**: The apt Ansible module (called with two parameters, name=apache2 and state=latest) checks whether the apache2 package is installed on the server, and if it isn't, it uses the apt-get tool to install it.

  - **Running the Apache2 service**: The service Ansible module (called with three parameters, name=apache2, state=started, and enabled=yes) checks whether the apache2 Unix service is started, and if it isn't, it uses the service command to start it.

Note that each task has a human-readable name, which is used in the console output, such that apt and service are Ansible modules, and name=apache2, state=latest, and state=started are module arguments. You already saw Ansible modules and arguments while using ad hoc commands. In the preceding playbook, we only defined one play, but there can be many of them, and each can be related to different groups of hosts.

> **Information**
>
> Note that since we used the apt Ansible module, the playbook is dedicated to Debian/Ubuntu servers.

For example, we could define two groups of servers in the inventory: `database` and `webservers`. Then, in the playbook, we could specify the tasks that should be executed on all database-hosting machines, and some different tasks that should be executed on all the web servers. By using one command, we could set up the whole environment.

Executing the playbook

When `playbook.yml` is defined, we can execute it using the `ansible-playbook` command:

```
$ ansible-playbook playbook.yml

PLAY [web1] *************************************************************
************

TASK [setup] ***********************************************************
************
ok: [web1]

TASK [ensure apache is at the latest version]
*****************************
changed: [web1]

TASK [ensure apache is running] ****************************************
************

ok: [web1]

PLAY RECAP *************************************************************
************
web1: ok=3 changed=1 unreachable=0 failed=0
```

> **Tip**
>
> If the server requires entering the password for the `sudo` command, then we need to add the `--ask-sudo-pass` option to the `ansible-playbook` command. It's also possible to pass the `sudo` password (if required) by setting the extra variable, `-e ansible_become_pass=<sudo_password>`.

The playbook configuration was executed, and therefore, the apache2 tool was installed and started. Note that if the task has changed something on the server, it is marked as changed. On the contrary, if there was no change, the task is marked as ok.

> **Tip**
> It is possible to run tasks in parallel by using the -f <num_of_threads> option.

The playbook's idempotency

We can execute the command again, as follows:

```
$ ansible-playbook playbook.yml

PLAY [web1] *********************************************
************

TASK [setup] *******************************************
************
ok: [web1]

TASK [ensure apache is at the latest version]
*****************************
ok: [web1]

TASK [ensure apache is running] ******************************
************
ok: [web1]

PLAY RECAP *********************************************
************
web1: ok=3 changed=0 unreachable=0 failed=0
```

Note that the output is slightly different. This time, the command didn't change anything on the server. That's because each Ansible module is designed to be idempotent. In other words, executing the same module many times in a sequence should have the same effect as executing it only once.

The simplest way to achieve idempotency is to always check first whether the task has been executed yet, and only execute it if it hasn't. Idempotency is a powerful feature, and we should always write our Ansible tasks this way.

If all the tasks are idempotent, then we can execute them as many times as we want. In that context, we can think of the playbook as a description of the desired state of remote machines. Then, the `ansible-playbook` command takes care of bringing the machine (or group of machines) into that state.

Handlers

Some operations should only be executed if some other tasks are changed. For example, imagine that you copy the configuration file to the remote machine and the Apache server should only be restarted if the configuration file has changed. *How could we approach such a case?*

Ansible provides an event-oriented mechanism to notify about the changes. In order to use it, we need to know two keywords:

- `handlers`: This specifies the tasks executed when notified.
- `notify`: This specifies the handlers that should be executed.

Let's look at the following example of how we could copy the configuration to the server and restart Apache only if the configuration has changed:

```
tasks:
- name: copy configuration
  copy:
    src: foo.conf
    dest: /etc/foo.conf
  notify:
  - restart apache
handlers:
- name: restart apache
  service:
    name: apache2
    state: restarted
```

Now, we can create the `foo.conf` file and run the `ansible-playbook` command:

```
$ touch foo.conf
$ ansible-playbook playbook.yml


...

TASK [copy configuration] ************************************
***********
changed: [web1]


RUNNING HANDLER [restart apache] *****************************
***********
changed: [web1]


PLAY RECAP **************************************************
***********
web1: ok=5 changed=2 unreachable=0 failed=0
```

> **Information**
>
> Handlers are always executed at the end of the play, and only once, even if triggered by multiple tasks.

Ansible copied the file and restarted the Apache server. It's important to understand that if we run the command again, nothing will happen. However, if we change the content of the `foo.conf` file and then run the `ansible-playbook` command, the file will be copied again (and the Apache server will be restarted):

```
$ echo "something" > foo.conf
$ ansible-playbook playbook.yml


...

TASK [copy configuration] ************************************
***********
changed: [web1]


RUNNING HANDLER [restart apache] *****************************
***********
```

```
changed: [web1]


PLAY RECAP *******************************************************
************
web1: ok=5 changed=2 unreachable=0 failed=0
```

We used the `copy` module, which is smart enough to detect whether the file has changed and then make a change on the server.

> **Tip**
> There is also a publish-subscribe mechanism in Ansible. Using it means assigning a topic to many handlers. Then, a task notifies the topic to execute all related handlers.

Variables

While the Ansible automation makes things identical and repeatable for multiple hosts, it is inevitable that servers may require some differences. For example, think of the application port number. It can be different, depending on the machine. Luckily, Ansible provides variables, which are a good mechanism to deal with server differences. Let's create a new playbook and define a variable:

```
---
- hosts: web1
  vars:
     http_port: 8080
```

The configuration defines the `http_port` variable with the value `8080`. Now, we can use it by using the `Jinja2` syntax:

```
tasks:
- name: print port number
  debug:
     msg: "Port number: {{ http_port }}"
```

> **Tip**
> The `Jinja2` language allows for doing way more than just getting a variable. We can use it to create conditions, loops, and much more. You can find more details on the Jinja page, at `https://jinja.palletsprojects.com/`.

The `debug` module prints the message while executing. If we run the `ansible-playbook` command, we can see the variable usage:

```
$ ansible-playbook playbook.yml

...

TASK [print port number] *****************************************
************
ok: [web1] => {
      "msg": "Port number: 8080"
}
```

Apart from user-defined variables, there are also predefined automatic variables. For example, the `hostvars` variable stores a map with the information regarding all hosts from the inventory. Using the Jinja2 syntax, we can iterate and print the IP addresses of all the hosts in the inventory:

```
---
- hosts: web1
  tasks:
  - name: print IP address
    debug:
      msg: "{% for host in groups['all'] %} {{
            hostvars[host]['ansible_host'] }} {% endfor %}"
```

Then, we can execute the `ansible-playbook` command:

```
$ ansible-playbook playbook.yml

...

TASK [print IP address] *****************************************
************
ok: [web1] => {
      "msg": " 192.168.64.12  192.168.64.13 "
}
```

Note that with the use of the Jinja2 language, we can specify the flow control operations inside the Ansible playbook file.

Roles

We can install any tool on the remote server by using Ansible playbooks. Imagine that we would like to have a server with MySQL. We could easily prepare a playbook similar to the one with the `apache2` package. However, if you think about it, a server with MySQL is quite a common case, and someone has surely already prepared a playbook for it, so maybe we can just reuse it. This is where Ansible roles and Ansible Galaxy come into play.

Understanding roles

An Ansible role is a well-structured playbook part prepared to be included in playbooks. Roles are separate units that always have the following directory structure:

```
templates/
tasks/
handlers/
vars/
defaults/
meta/
```

> **Information**
>
> You can read more about roles and what each directory means on the official Ansible page at `https://docs.ansible.com/ansible/latest/user_guide/playbooks_reuse_roles.html`.

In each of the directories, we can define the `main.yml` file, which contains the playbook parts that can be included in the `playbook.yml` file. Continuing the MySQL case, there is a role defined on GitHub at `https://github.com/geerlingguy/ansible-role-mysql`. This repository contains task templates that can be used in our playbook. Let's look at a part of the `tasks/setup-Debian.yml` file, which installs the `mysql` package in Ubuntu/Debian:

```
...
- name: Ensure MySQL Python libraries are installed.
  apt:
    name: "{{ mysql_python_package_debian }}"
    state: present


- name: Ensure MySQL packages are installed.
  apt:
```

```
    name: "{{ mysql_packages }}"
    state: present
    register: deb_mysql_install_packages
...
```

This is only one of the tasks defined in the `tasks/main.yml` file. Others tasks are responsible for the installation of MySQL into other operating systems.

If we use this role in order to install MySQL on the server, it's enough to create the following `playbook.yml`:

```
---
- hosts: all
  become: yes
  become_method: sudo
  roles:
  - role: geerlingguy.mysql
```

Such a configuration installs the MySQL database on all servers using the `geerlingguy.mysql` role.

Ansible Galaxy

Ansible Galaxy is to Ansible what Docker Hub is to Docker—it stores common roles so that they can be reused by others. You can browse the available roles on the Ansible Galaxy page at `https://galaxy.ansible.com/`.

To install a role from Ansible Galaxy, we can use the `ansible-galaxy` command:

```
$ ansible-galaxy install username.role_name
```

This command automatically downloads the role. In the case of the MySQL example, we could download the role by executing the following:

```
$ ansible-galaxy install geerlingguy.mysql
```

The command downloads the `mysql` role, which can later be used in the playbook file. If you defined `playbook.yml` as described in the preceding snippet, the following command installs MySQL into all of your servers:

```
$ ansible-playbook playbook.yml
```

Now that you know about the basics of Ansible, let's see how we can use it to deploy our own applications.

Deployment with Ansible

We have covered the most fundamental features of Ansible. Now, let's forget, just for a little while, about Docker, Kubernetes, and most of the things we've learned so far. Let's configure a complete deployment step by only using Ansible. We will run the calculator service on one server and the Hazelcast service on the second server.

Installing Hazelcast

We can specify a play in the new playbook. Let's create the `playbook.yml` file, with the following content:

```
---
- hosts: web1
  become: yes
  become_method: sudo
  tasks:
  - name: ensure Java Runtime Environment is installed
    apt:
      name: default-jre
      state: present
      update_cache: yes
  - name: create Hazelcast directory
    file:
      path: /var/hazelcast
      state: directory
  - name: download Hazelcast
    get_url:
      url: https://repo1.maven.org/maven2/com/hazelcast/
hazelcast/5.0.2/hazelcast-5.0.2.jar
      dest: /var/hazelcast/hazelcast.jar
      mode: a+r
  - name: copy Hazelcast starting script
    copy:
      src: hazelcast.sh
```

```
      dest: /var/hazelcast/hazelcast.sh
      mode: a+x
  - name: configure Hazelcast as a service
    file:
      path: /etc/init.d/hazelcast
      state: link
      force: yes
      src: /var/hazelcast/hazelcast.sh
  - name: start Hazelcast
    service:
      name: hazelcast
      enabled: yes
      state: started
```

The configuration is executed on the web1 server and it requires root permissions. It performs a few steps that will lead to a complete Hazelcast server installation. Let's walk through what we defined:

1. **Prepare the environment**: This task ensures that the Java runtime environment is installed. Basically, it prepares the server environment so that Hazelcast will have all the necessary dependencies. With more complex applications, the list of dependent tools and libraries can be way longer.

2. **Download Hazelcast tool**: Hazelcast is provided in the form of a JAR, which can be downloaded from the internet. We hardcoded the version, but in a real-life scenario, it would be better to extract it to a variable.

3. **Configure application as a service**: We would like to have Hazelcast running as a Unix service so that it would be manageable in the standard way. In this case, it's enough to copy a service script and link it in the /etc/init.d/ directory.

4. **Start the Hazelcast service**: When Hazelcast is configured as a Unix service, we can start it in the standard way.

In the same directory, let's create hazelcast.sh, which is a script (shown as follows) that is responsible for running Hazelcast as a Unix service:

```
#!/bin/bash


### BEGIN INIT INFO
# Provides: hazelcast
# Required-Start: $remote_fs $syslog
```

```
# Required-Stop: $remote_fs $syslog
# Default-Start: 2 3 4 5
# Default-Stop: 0 1 6
# Short-Description: Hazelcast server
### END INIT INFO


java -cp /var/hazelcast/hazelcast.jar com.hazelcast.core.
server.HazelcastMemberStarter &
```

After this step, we could execute the playbook and have Hazelcast started on the web1 server machine. However, let's first create a second play to start the calculator service, and then run it all together.

Deploying a web service

We prepare the calculator web service in two steps:

1. Change the Hazelcast host address.

2. Add calculator deployment to the playbook.

Changing the Hazelcast host address

Previously, we hardcoded the Hazelcast host address as hazelcast, so now we should change it in the src/main/java/com/leszko/calculator/ CalculatorApplication.java file to 192.168.64.12 (the same IP address we have in our inventory, as web1).

> **Tip**
> In real-life projects, the application properties are usually kept in the properties file. For example, for the Spring Boot framework, it's a file called application.properties or application.yml. Then, we could change them with Ansible and therefore be more flexible.

Adding calculator deployment to the playbook

Finally, we can add the deployment configuration as a new play in the playbook. yml file. It is similar to the one we created for Hazelcast:

```
- hosts: web2
  become: yes
```

```yaml
  become_method: sudo
  tasks:
  - name: ensure Java Runtime Environment is installed
    apt:
      name: default-jre
      state: present
      update_cache: yes
  - name: create directory for Calculator
    file:
      path: /var/calculator
      state: directory
  - name: copy Calculator starting script
    copy:
      src: calculator.sh
      dest: /var/calculator/calculator.sh
      mode: a+x
  - name: configure Calculator as a service
    file:
      path: /etc/init.d/calculator
      state: link
      force: yes
      src: /var/calculator/calculator.sh
  - name: copy Calculator
    copy:
      src: build/libs/calculator-0.0.1-SNAPSHOT.jar
      dest: /var/calculator/calculator.jar
      mode: a+x
    notify:
    - restart Calculator
  handlers:
  - name: restart Calculator
    service:
      name: calculator
      enabled: yes
      state: restarted
```

The configuration is very similar to what we saw in the case of Hazelcast. One difference is that this time, we don't download the JAR from the internet, but we copy it from our filesystem. The other difference is that we restart the service using the Ansible handler. That's because we want to restart the calculator each time a new version is copied.

Before we start it all together, we also need to define `calculator.sh`:

```bash
#!/bin/bash


### BEGIN INIT INFO
# Provides: calculator
# Required-Start: $remote_fs $syslog
# Required-Stop: $remote_fs $syslog
# Default-Start: 2 3 4 5
# Default-Stop: 0 1 6
# Short-Description: Calculator application
### END INIT INFO


java -jar /var/calculator/calculator.jar &
```

When everything is prepared, we will use this configuration to start the complete system.

Running the deployment

As always, we can execute the playbook using the `ansible-playbook` command. Before that, we need to build the calculator project with Gradle:

```
$ ./gradlew build
$ ansible-playbook playbook.yml
```

After the successful deployment, the service should be available, and we can check that it's working at `http://192.168.64.13:8080/sum?a=1&b=2` (the IP address should be the same one that we have in our inventory as web2). As expected, it should return 3 as the output.

Note that we have configured the whole environment by executing one command. What's more, if we need to scale the service, then it's enough to add a new server to the inventory and rerun the `ansible-playbook` command. Also, note that we could package it as an Ansible role and upload it to GitHub, and from then on, everyone could run the same system on their Ubuntu servers. That's the power of Ansible!

We have shown how to use Ansible for environmental configuration and application deployment. The next step is to use Ansible with Docker and Kubernetes.

Ansible with Docker and Kubernetes

As you may have noticed, Ansible and Docker (along with Kubernetes) address similar software deployment issues:

- **Environmental configuration**: Both Ansible and Docker provide a way to configure the environment; however, they use different means. While Ansible uses scripts (encapsulated inside the Ansible modules), Docker encapsulates the whole environment inside a container.

- **Dependencies**: Ansible provides a way to deploy different services on the same or different hosts and lets them be deployed together. Kubernetes has similar functionality, which allows for running multiple containers at the same time.

- **Scalability**: Ansible helps to scale the services providing the inventory and host groups. Kubernetes has similar functionality to automatically increase or decrease the number of running containers.

- **Automation with configuration files**: Docker, Kubernetes, and Ansible store the whole environmental configuration and service dependencies in files (stored in the source control repository). For Ansible, this file is called `playbook.yml`. In the case of Docker and Kubernetes, we have `Dockerfile` for the environment and `deployment.yml` for the dependencies and scaling.

- **Simplicity**: Both tools are very simple to use and provide a way to set up the whole running environment with a configuration file and just one command execution.

If we compare the tools, Docker does a little more, since it provides isolation, portability, and a kind of security. We could even imagine using Docker/Kubernetes without any other configuration management tools. Then, *why do we need Ansible at all?*

Benefits of Ansible

Ansible may seem redundant; however, it brings additional benefits to the delivery process, which are as follows:

- **Docker environment**: The Docker/Kubernetes hosts themselves have to be configured and managed. Every container is ultimately running on Linux machines, which need kernel patching, Docker Engine updates, and network configuration, for example. What's more, there may be different server machines with different Linux distributions, and the responsibility of Ansible is to make sure everything is up and running.

- **Non-Dockerized applications**: Not everything is run inside a container. If part of the infrastructure is containerized and part is deployed in the standard way or in the cloud, then Ansible can manage it all with the playbook configuration file. There may be different reasons for not running an application as a container; for example, performance, security, specific hardware requirements, or working with the legacy software.

- **Inventory**: Ansible offers a very friendly way to manage the physical infrastructure by using inventories, which store information about all the servers. It can also split the physical infrastructure into different environments—production, testing, and development.

- **Cloud provisioning**: Ansible can be responsible for provisioning Kubernetes clusters or installing Kubernetes in the cloud; for example, we can imagine integration tests in which the first step is to create a Kubernetes cluster on **Google Cloud Platform** (**GCP**) (only then can we deploy the whole application and perform the testing process).

- **GUI**: Ansible offers GUI managers (commercial Ansible Tower and open source AWX), which aim to improve the experience of infrastructure management.

- **Improving the testing process**: Ansible can help with integration and acceptance testing, as it can encapsulate testing scripts.

We can look at Ansible as the tool that takes care of the infrastructure, while Docker and Kubernetes are tools that take care of the environmental configuration and clustering. An overview is presented in the following diagram:

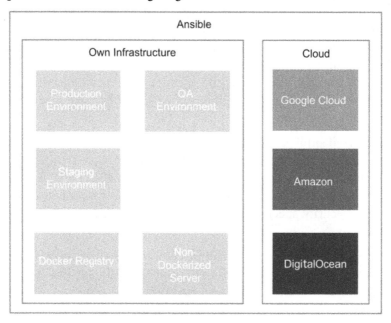

Figure 7.2 – Ansible as the infrastructure manager

Ansible manages the infrastructure: **Kubernetes clusters**, **Docker servers**, **Docker registries**, **servers without Docker**, and **cloud providers**. It also takes care of the physical location of the servers. Using the inventory host groups, it can link the web services to the databases that are close to their geographic locations.

Let's look at how we can use Ansible to install Docker on a server and deploy a sample application.

The Ansible Docker playbook

Ansible integrates with Docker smoothly, because it provides a set of Docker-dedicated modules. If we create an Ansible playbook for Docker-based deployment, then the first task is to make sure that the Docker Engine is installed on every machine. Then, it should run a container using Docker.

First, let's install Docker on an Ubuntu server.

Installing Docker

We can install the Docker Engine by using the following task in the Ansible playbook:

```yaml
- hosts: web1
  become: yes
  become_method: sudo
  tasks:
  - name: Install required packages
    apt:
      name: "{{ item }}"
      state: latest
      update_cache: yes
    loop:
    - apt-transport-https
    - ca-certificates
    - curl
    - software-properties-common
    - python3-pip
    - virtualenv
    - python3-setuptools
  - name: Add Docker GPG apt Key
    apt_key:
      url: https://download.docker.com/linux/ubuntu/gpg
      state: present
  - name: Add Docker Repository
    apt_repository:
      repo: deb https://download.docker.com/linux/ubuntu focal
stable
      state: present
  - name: Update apt and install docker-ce
    apt:
      name: docker-ce
      state: latest
      update_cache: yes
  - name: Install Docker Module for Python
    pip:
      name: docker
```

> **Information**
>
> The playbook looks slightly different for each operating system. The one presented here is for Ubuntu 20.04.

This configuration installs Docker and Docker Python tools (needed by Ansible). Note that we used a new Ansible syntax, `loop`, in order to make the playbook more concise.

When Docker is installed, we can add a task that will run a Docker container.

Running Docker containers

Running Docker containers is done with the use of the `docker_container` module, and it looks as follows:

```yaml
- hosts: web1
  become: yes
  become_method: sudo
  tasks:
  - name: run Hazelcast container
    community.docker.docker_container:
      name: hazelcast
      image: hazelcast/hazelcast
      state: started
      exposed_ports:
       - 5701
```

> **Information**
>
> You can read more about all of the options of the `docker_container` module at `https://docs.ansible.com/ansible/latest/collections/community/docker/docker_container_module.html`.

With the two playbooks presented previously, we configured the Hazelcast server using Docker. Note that this is very convenient because we can run the same playbook on multiple (Ubuntu) servers.

Now, let's take a look at how Ansible can help with Kubernetes.

The Ansible Kubernetes playbook

Similar to Docker, Ansible can help with Kubernetes. When you have your Kubernetes cluster configured, then you can create Kubernetes resources using the Ansible k8s module. Here's a sample Ansible task to create a namespace in Kubernetes:

```
- name: Create namespace
  kubernetes.core.k8s:
    name: my-namespace
    api_version: v1
    kind: Namespace
    state: present
```

The configuration here makes sure a namespace called my-namespace is created in the Kubernetes cluster.

> **Information**
>
> You can find more information about the Ansible k8s module at https://docs.ansible.com/ansible/latest/collections/kubernetes/core/k8s_module.html.

We have covered configuration management with Ansible, which is a perfect approach if your deployment environment consists of bare-metal servers. You can also use Ansible with cloud providers, and there are a number of modules dedicated to that purpose. For example, amazon.aws.ec2_instance lets you create and manage AWS EC2 instances. However, when it comes to the cloud, there are better solutions. Let's see what they are and how to use them.

Introducing IaC

IaC is the process of managing and provisioning computing resources instead of physical hardware configuration. It is mostly associated with the cloud approach, in which you can request the necessary infrastructure in a programmable manner.

Managing computer infrastructure was always a hard, time-consuming, and error-prone activity. You had to manually place the hardware, connect the network, install the operating system, and take care of its updates. Together with the cloud, things became simple; all you had to do was to write a few commands or make a few clicks in the web UI. IaC goes one step further, as it allows you to specify in a declarative manner what infrastructure you need. To understand it better, let's take a look at the following diagram:

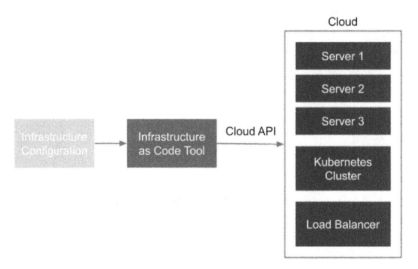

Figure 7.3 – IaC

You prepare a declarative description of your infrastructure, for example, that you need three servers, a Kubernetes cluster, and a load balancer. Then, you pass this configuration to a tool that uses a cloud-specific API (for example, the AWS API) in order to make sure the infrastructure is as requested. Note that you should store the infrastructure configuration in the source code repository, and you can create multiple identical environments from the same configuration.

You can see that the IaC idea is very similar to configuration management; however, while configuration management makes sure your software is configured as specified, IaC makes sure that your infrastructure is configured as specified.

Now, let's look into the benefits of using IaC.

Benefits of IaC

There are a number of benefits that infrastructure brings into all DevOps activities. Let's walk through the most important ones:

- **Speed**: Creating the whole infrastructure means nothing more than running a script, which significantly reduces the time needed before we can start deploying the applications.

- **Cost reduction**: Automating the infrastructure provisioning reduces the number of DevOps team members required to operate server environments.

- **Consistency**: IaC configuration files become the single point of truth, so they guarantee that every created environment is exactly the same.

- **Risk reduction**: Infrastructure configuration is stored in the source code repository and follows the standard code review process, which reduces the probability of making a mistake.

- **Collaboration**: Multiple people can share the code and work on the same configuration files, which increases work efficiency.

I hope these points have convinced you that IaC is a great approach. Let's now look into the tools you can use for IaC.

Tools for IaC

When it comes to IaC, there are a number of tools you can use. The choice depends on the cloud provider you use and on your own preferences. Let's walk through the most popular solutions:

- **Terraform**: The most popular IaC tool on the market. It's open source and uses plugin-based modules called *providers* to support different infrastructure APIs. Currently, more than 1,000 Terraform providers exist, including AWS, Azure, GCP, and DigitalOcean.

- **Cloud provider specific**: Each major cloud provider has its own IaC tool:

  - **AWS CloudFormation**: An Amazon service that allows you to specify AWS resources in the form of YAML or JSON template files

  - **Azure Resource Manager (ARM)**: A Microsoft Azure service that allows you to create and manage Azure resources with the use of ARM template files

  - **Google Cloud Deployment Manager**: A Google service that allows you to manage Google Cloud Platform resources with the use of YAML files

- **General configuration management**: Ansible, Chef, and Puppet all provide dedicated modules to provision the infrastructure in the most popular cloud solutions.

- **Pulumi**: A very flexible tool that allows you to specify the desired infrastructure using general-purpose programming languages, such as JavaScript, Python, Go, or C#.

- **Vagrant**: Usually associated with virtual machine management, it provides a number of plugins to provision infrastructure using AWS and other cloud providers.

Of all the solutions mentioned, Terraform is by far the most popular. That is why we'll spend some more time understanding how it works.

Introduction to Terraform

Terraform is an open source tool created and maintained by HashiCorp. It allows you to specify your infrastructure in the form of human-readable configuration files. Similar to Ansible, it works in a declarative manner, which means that you specify the expected outcome, and Terraform makes sure your environment is created as specified.

Before we dive into a concrete example, let's spend a moment understanding how Terraform works.

Understanding Terraform

Terraform reads a configuration file and adjusts the cloud resources accordingly. Let's look at the following diagram, which presents this process:

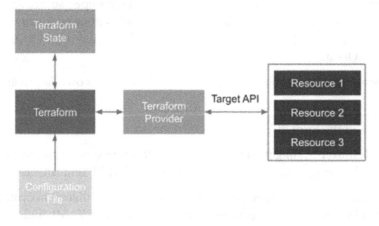

Figure 7.4 – Terraform workflow

A user creates **Configuration File** and starts the **Terraform** tool. Then, **Terraform** checks the **Terraform State** and uses **Terraform Provider** to translate the declarative configuration file into the requests called against **Target API,** which is specific for the given cloud provider. As an example, we can think of a configuration file that defines three AWS EC2 instances. Terraform uses the AWS provider, which executes requests to the AWS API to make sure that three AWS EC2 instances are created.

Information

There are more than 1,000 Terraform providers available, and you can browse them via the Terraform Registry at `https://registry.terraform.io/`.

The Terraform workflow always consists of three stages:

- **Write**: User defines cloud resources as a configuration file.

- **Plan**: Terraform compares the configuration file with the current state and prepares the execution plan.

- **Apply**: User approves the plan and Terraform executes the planned operations using the cloud API.

This approach is very convenient because, with the plan stage, we can always check what Terraform is going to change in our infrastructure, before actually applying the change.

Now that we understand the idea behind Terraform, let's look at how it all works in practice, starting from the Terraform installation process.

Installing Terraform

The installation process depends on the operating system. In the case of Ubuntu, you can execute the following commands:

```
$ curl -fsSL https://apt.releases.hashicorp.com/gpg | sudo
apt-key add -
$ sudo apt-add-repository "deb [arch=amd64] https://apt.
releases.hashicorp.com $(lsb_release -cs) main"
$ sudo apt-get update
$ sudo apt-get install terraform
```

> **Information**
>
> You can find the installation guides for all the operating systems on the official Terraform website, at `https://www.terraform.io/downloads`.

After the installation process, we can verify that the `terraform` command works correctly:

```
$ terraform version
Terraform v1.1.5
```

After Terraform is configured, we can move to the Terraform example.

Using Terraform

As an example, let's use Terraform to provision an AWS EC2 instance. For this purpose, we need to first configure AWS.

Configuring AWS

To access AWS from your machine, you will need the following:

- An AWS account
- The AWS CLI installed

> **Information**
>
> You can create a free AWS account at `https://aws.amazon.com/free`. To install the AWS CLI tool, please check the following instructions: `https://docs.aws.amazon.com/cli/latest/userguide/getting-started-install.html`.

Let's configure the AWS CLI with the following command:

```
$ aws configure
```

The AWS command prompts your AWS access key ID and AWS secret access key.

> **Information**
>
> For instructions on how to create an AWS access key pair, please visit `https://docs.aws.amazon.com/cli/latest/userguide/cli-configure-quickstart.html#cli-configure-quickstart-creds`.

After these steps, access to your AWS account is configured and we can start playing with Terraform.

Writing Terraform configuration

In a fresh directory, let's create the `main.tf` file and add the following content:

```
terraform {
  required_version = ">= 1.1"                   (1)
  required_providers {
    aws = {                                      (2)
      source  = "hashicorp/aws"
      version = "~> 3.74"
    }
  }
}
provider "aws" {
  profile = "default"                            (3)
  region  = "us-east-1"                          (4)
}
resource "aws_instance" "my_instance" {          (5)
  ami             = "ami-04505e74c0741db8d"      (6)
  instance_type = "t2.micro"                     (7)
}
```

In the preceding configuration, we defined the following parts:

1. The Terraform tool version should be at least `1.1`.

2. The configuration uses the `hashicorp/aws` provider:

    -   The provider version needs to be at least `3.74`.

    -   Terraform will automatically download it from the **Terraform Registry**.

3. The credentials for the `aws` provider are stored in the `default` location created by the AWS CLI.

4. The provider creates all resources in the `us-east-1` region.

5. The provider creates `aws_instance` (an AWS EC2 instance) named `my_instance`.

6. An EC2 instance is created from `ami-04505e74c0741db8d` (Ubuntu 20.04 LTS in the `us-east-1` region).

7. The instance type is `t2.micro`.

You can see that the whole configuration is declarative. In other words, we define what we want, not the algorithm for how to achieve it.

When the configuration is created, we need to download the required provider from the Terraform Registry.

Initializing Terraform configuration

Let's execute the following command:

```
$ terraform init
```

This command downloads all required providers and stores them in the `.terraform` directory. Now, let's finally apply the Terraform configuration.

Applying Terraform configuration

Before we make any Terraform changes, it's good to first execute `terraform plan` to check what changes stand ahead of us:

```
$ terraform plan
Terraform used the selected providers to generate the following
execution plan. Resource actions are indicated with the
following symbols:
  + create
...
```

We can see that by applying the configuration, we will create a resource in our infrastructure as described in the console output.

Let's now apply our configuration:

```
$ terraform apply
...
Do you want to perform these actions?
  Terraform will perform the actions described above.
```

```
Only 'yes' will be accepted to approve.
Enter a value: yes
...
Apply complete! Resources: 1 added, 0 changed, 0 destroyed.
```

After confirming the change, you should see a lot of logs and the last `Apply complete!` message, which means that our infrastructure is created.

Now, let's verify that everything is as expected.

Verifying the infrastructure

From the Terraform perspective, we can execute the following command to see the state of our infrastructure:

```
$ terraform show
# aws_instance.my_instance:
resource "aws_instance" "my_instance" {
...
}
```

This prints all the information about the resource we created.

> **Information**
> Terraform, the same as Ansible, favors idempotent operations. That is why, if we execute `terraform plan` or `terraform apply` again, nothing will change. You will only see the following message: `No changes. Your infrastructure matches the configuration.`

We can now verify that our AWS EC2 instance is really created. Since we already installed the AWS CLI, we can check it with the following command:

```
$ aws ec2 describe-instances --region us-east-1
{
    "Reservations": [
        {
            "Groups": [],
            "Instances": [
                {
                    "AmiLaunchIndex": 0,
```

```
            "ImageId": "ami-04505e74c0741db8d",
            "InstanceId": "i-053b633c810728a97",
            "InstanceType": "t2.micro",
...
```

If you prefer, you can also check in the AWS web console that the instance is created.

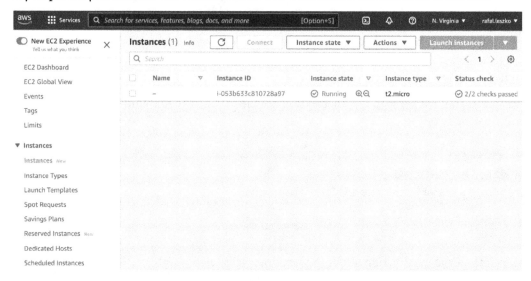

Figure 7.5 – AWS EC2 instance created with Terraform

We just verified that our Terraform configuration works as expected.

> **Tip**
>
> When working together with Ansible, we can make use of Ansible's dynamic inventories and let Ansible discover created EC2 instances. Read more at `https://docs.ansible.com/ansible/latest/user_guide/intro_dynamic_inventory.html`.

To make our example complete, let's also see how to delete created resources.

Destroying the infrastructure

Let's remove the resources we created with the following command:

```
$ terraform destroy
aws_instance.my_instance: Refreshing state... [id=i-
053b633c810728a97]
```

```
Terraform used the selected providers to generate the following
execution plan. Resource actions are indicated with the
following symbols:
  - destroy
...
Do you really want to destroy all resources?
  Terraform will destroy all your managed infrastructure, as
shown above.
  There is no undo. Only 'yes' will be accepted to confirm.
  Enter a value: yes
...
Destroy complete! Resources: 1 destroyed.
```

After the user confirmation, Terraform removed all the resources. You can check that our AWS EC2 instance does not exist anymore.

As the last thing with Terraform, let's see how it interacts with Kubernetes.

Terraform and Kubernetes

There are two different use cases when it comes to the interaction between Terraform and Kubernetes:

- Provisioning a Kubernetes cluster
- Interacting with a Kubernetes cluster

Let's present them one by one.

Provisioning a Kubernetes cluster

Each of the major cloud providers offers managed Kubernetes clusters, and we can provision them using Terraform. The following Terraform providers are available:

- **AWS**: This can provision clusters in Amazon **Elastic Kubernetes Service (EKS)**.
- **Google**: This can provision clusters in **Google Kubernetes Engine (GKE)**.
- **AzureRM**: This can provision clusters in **Azure Kubernetes Service (AKS)**.

Using each of these providers is relatively simple and works similarly to how we described in our Terraform example.

> **Tip**
>
> If you install Kubernetes on bare-metal servers, you should use a configuration management tool, such as Ansible. To provision a cloud-managed Kubernetes cluster, you can use either Ansible or Terraform, but the former is a better fit.

Let's also look at the second usage of Terraform with Kubernetes.

Interacting with a Kubernetes cluster

Similar to Ansible, we can use Terraform to interact with a Kubernetes cluster. In other words, instead of applying Kubernetes configurations using the `kubectl` command, we can use a dedicated Terraform Kubernetes provider.

A sample Terraform configuration to change Kubernetes resources looks as follows:

```
resource "kubernetes_namespace" "example" {
  metadata {
    name = "my-first-namespace"
  }
}
```

The preceding configuration creates a namespace called `my-namespace` in the Kubernetes cluster.

> **Tip**
>
> There are multiple ways you can interact with a Kubernetes cluster: `kubectl`, Ansible, Terraform, or some other tool. As a rule of thumb, I would always first try the simplest approach, which is the `kubectl` command, and only incorporate Ansible or Terraform if you have some special requirements; for example, you manage multiple Kubernetes clusters at the same time.

We covered the basics of Terraform, so let's wrap up this chapter with a short summary.

Summary

We have covered configuration management and IaC approaches, together with the related tooling. Note that whether you should use Ansible, Terraform, or neither of them inside your continuous delivery pipeline highly depends on your particular use case.

Ansible shines when you have multiple bare-metal servers to manage, so if your release means making the same change into many servers at the same time, you'll most probably place Ansible commands inside your pipeline.

Terraform works best when you use the cloud. Therefore, if your release means making a change to your cloud infrastructure, then Terraform is the way to go.

However, if your environment is only a single Kubernetes cluster, then there is nothing wrong with executing `kubectl` commands inside your pipeline.

The other takeaway points from this chapter are as follows:

- Configuration management is the process of creating and applying the configurations of the application.
- Ansible is one of the most trending configuration management tools. It is agentless, and therefore, it requires no special server configuration.
- Ansible can be used with ad hoc commands, but the real power lies in Ansible playbooks.
- The Ansible playbook is a definition of how the environment should be configured.
- The purpose of Ansible roles is to reuse parts of playbooks.
- Ansible Galaxy is an online service to share Ansible roles.
- IaC is a process of managing cloud resources.
- Terraform is the most popular tool for IaC.

In the next chapter, we will wrap up the continuous delivery process and complete the final Jenkins pipeline.

Exercises

In this chapter, we covered the fundamentals of Ansible and ways to use it with Docker and Kubernetes. As exercises, try the following tasks:

1. Create the server infrastructure and use Ansible to manage it:

    I. Connect a physical machine or run a VirtualBox machine to emulate the remote server.

    II. Configure SSH access to the remote machine (SSH keys).

    III. Install Python on the remote machine.

IV. Create an Ansible inventory with the remote machine.

V. Run the Ansible ad hoc command (with the `ping` module) to check that the infrastructure is configured correctly.

2. Create a Python-based `hello world` web service and deploy it in a remote machine using Ansible playbook:

I. The service can look exactly the same as we described in the exercises for the chapter.

II. Create a playbook that deploys the service into the remote machine.

III. Run the `ansible-playbook` command and check whether the service was deployed.

3. Provision a GCP virtual machine instance using Terraform:

I. Create an account in GCP.

II. Install the `gcloud` tool and authenticate (`gcloud init`).

III. Generate credentials and export them into the `GOOGLE_APPLICATION_CREDENTIALS` environment variable.

IV. Create a Terraform configuration that provisions a virtual machine instance.

V. Apply the configuration using Terraform.

VI. Verify that the instance was created.

Questions

To verify your knowledge from this chapter, please answer the following questions:

1. What is configuration management?
2. What does it mean that the configuration management tool is agentless?
3. What are the three most popular configuration management tools?
4. What is Ansible inventory?
5. What is the difference between Ansible ad hoc commands and playbooks?
6. What is an Ansible role?
7. What is Ansible Galaxy?
8. What is IaC?
9. What are the most popular tools for IaC?

Further reading

To read more about configuration management and IaC, please refer to the following resources:

- **Official Ansible documentation**: `https://docs.ansible.com/`

- **Official Terraform documentation**: `https://www.terraform.io/docs`

- **Michael T. Nygard, Release It!**: (`https://pragprog.com/titles/mnee2/release-it-second-edition/`)

- **Russ McKendrick, Learn Ansible**: (`https://www.packtpub.com/virtualization-and-cloud/learn-ansible`)

8

Continuous Delivery Pipeline

In this chapter, we will focus on the missing parts of the final pipeline, which are the environments and infrastructure, application versioning, and non-functional testing.

We will be covering the following topics:

- Environments and infrastructure
- Non-functional testing
- Application versioning
- Completing the continuous delivery pipeline

Technical requirements

To follow this chapter, you'll need the following:

- A Jenkins instance (with Java 8+, Docker, and `kubectl` installed on your Jenkins agents)
- A Docker registry (for example, an account on Docker Hub)
- Two Kubernetes clusters

All the examples and solutions for the exercises in this chapter can be found on GitHub at `https://github.com/PacktPublishing/Continuous-Delivery-With-Docker-and-Jenkins-3rd-Edition/tree/main/Chapter08`.

Code in Action videos for this chapter can be viewed at `https://bit.ly/3JeyQ1X`.

Environments and infrastructure

So far, we have deployed our applications to some servers – that is, Docker hosts, Kubernetes clusters, and pure Ubuntu servers (in the case of Ansible). However, when we think more deeply about the **continuous delivery (CD)** process (or the software delivery process in general), we need to logically group our resources. There are two main reasons why this is important:

- The physical location of machines matters
- No testing should be done on the production machines

Taking these facts into consideration, in this section, we will discuss different types of environments, their role in the CD process, and the security aspect of our infrastructure.

Types of environments

There are four common environment types – **production**, **staging**, **QA** (testing), and **development**. Let's discuss each of them one by one.

Production

The production environment is the environment that is used by the end user. It exists in every company and is the most important environment.

The following diagram shows how most production environments are organized:

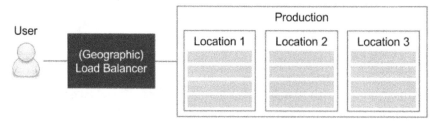

Figure 8.1 – Production environment

Users access the service through the load balancer, which chooses the machine. If the application is released in multiple physical locations, then the (first) device is usually a DNS-based geographic load balancer. In each location, we have a cluster of servers. If we use Docker and Kubernetes, for example, this means that in each location, we have at least one Kubernetes cluster.

The physical location of machines matters because the request-response time can differ significantly, depending on the physical distance. Moreover, the database and other dependent services should be located on a machine that is close to where the service is deployed. What's even more important is that the database should be sharded in a way that minimizes the replication overhead between different locations; otherwise, we may end up waiting for the databases to reach a consensus between their instances, which will be located far away from each other. More details on the physical aspects are beyond the scope of this book, but it's important to remember that Docker and Kubernetes themselves do not solve this problem.

> **Information**
> Containerization and virtualization allow you to think about servers as infinite resources; however, some physical aspects such as location are still relevant.

Staging

The staging environment is where the release candidate is deployed to perform the final tests before going live. Ideally, this environment is a mirror of the production environment.

The following diagram shows what such an environment should look like in the context of the delivery process:

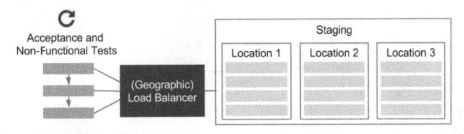

Figure 8.2 – Staging environment

Note that the staging environment is a clone of the production environment. If the application is deployed in multiple locations, then the staging environment should also have multiple locations.

In the CD process, all automated acceptance tests (both functional and non-functional) are run against this environment. While most functional tests don't usually require identical production-like infrastructure, in the case of non-functional (especially performance) tests, it is a must.

To save costs, it's not uncommon for the staging infrastructure to differ from the production environment (usually, it contains fewer machines). Such an approach can, however, lead to many production issues. *Michael T. Nygard, in Release It! Design and Deploy Production-Ready Software*, gives an example of a real-life scenario in which fewer machines were used in the staging environment than in production.

The story goes like this: in one company, the system was stable until a certain code change caused the production environment to become extremely slow, even though all the stress tests were passed. *How was this possible?* This happened because there was a synchronization point where each server communicated with the others. In the case of the staging environment, there was one server, so there was no blocker. In production, however, there were many servers, which resulted in servers waiting for each other. This example is just the tip of the iceberg, and many production issues may fail to be tested by acceptance tests if the staging environment is different from the production environment.

QA

The QA environment (also called the testing environment) is intended for the QA team to perform exploratory testing and for external applications (that depend on our service) to perform integration testing. The use cases and the infrastructure of the QA environment are shown in the following diagram:

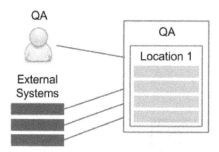

Figure 8.3 – QA environment

While staging does not need to be stable (in the case of CD, it is changed after every code change that's committed to the repository), the QA instance needs to provide a certain amount of stability and expose the same (or backward-compatible) API as the production environment. In contrast to the staging environment, the infrastructure can be different from that of the production environment since its purpose is not to ensure that the release candidate works properly.

A very common case is to allocate fewer machines (for example, only from one location) to the QA instance.

> **Information**
>
> Deploying to the QA environment is usually done in a separate pipeline so that it's independent of the automatic release process. Such an approach is convenient because the QA instance has a different life cycle than the production one (for instance, the QA team may want to perform testing on the experimental code branched from the trunk).

Development

The development environment can be created as a shared server for all developers, or each developer can have a development environment for themselves. The following is a simple diagram of this:

Figure 8.4 – Development environment

The development environment always contains the latest version of the code. It is used to enable integration between developers and can be treated the same way as the QA environment. However, it is used by developers, not QAs.

Now that we've looked at all the environments, let's see how they fit into the CD process.

Environments in continuous delivery

In the CD process, the staging environment is indispensable. In some very rare cases, when performance is not important and the project doesn't have many dependencies, we can perform the acceptance tests on the local (development) Docker host, but that should be an exception, not a rule. In such cases, we always risk some production issues occurring that are related to the environment.

The other environments are usually not important in terms of CD. If we would like to deploy to the QA or development environment with every commit, then we can create separate pipelines for that purpose (being careful not to obscure the main release pipeline). In many cases, deployment to the QA environment is triggered manually since it has a different life cycle from production.

Securing environments

All environments need to be well secured – that's clear. What's even more obvious is that the most important requirement is to keep the production environment secure because our business depends on it, and the consequences of any security flaw can be the most serious.

> **Information**
>
> Security is a broad topic. In this section, we will only focus on the topics related to the CD process. Nevertheless, setting up a complete server infrastructure requires much more security knowledge.

In the CD process, the Jenkins agent must have access to servers so that it can deploy the application.

There are different approaches for providing agents with the server's credentials:

- **Put an SSH key in the agent**: If we don't use dynamic Docker slave provisioning, then we can configure Jenkins agent machines so that they contain private SSH keys.

- **Put an SSH key in the agent image**: If we use dynamic Docker slave provisioning, we can add the SSH private key to the Docker agent image; however, this creates a possible security hole since anyone who has access to that image would have access to the production servers.

- **Use Jenkins credentials**: We can configure Jenkins to store credentials and use them in the pipeline.

- **Copy to the slave Jenkins plugin**: We can copy the SSH key dynamically into the slave while starting the Jenkins build.

Each solution has some advantages and drawbacks. While using any of them, we have to take extra caution since, when an agent has access to the production environment, anyone breaking into that agent can break into the production environment.

The riskiest solution is to put SSH private keys into the Jenkins agent image since everywhere the image is stored (the Docker registry or Docker host within Jenkins) needs to be well secured.

Now that we've covered the infrastructure, let's look at a topic that we haven't covered yet – non-functional testing.

Non-functional testing

We learned a lot about functional requirements and automated acceptance testing in the previous chapters. *But what should we do with non-functional requirements? Or even more challenging, what if there are no requirements? Should we skip them in the CD process?* We will answer these questions throughout this section.

Non-functional aspects of the software are always important because they can cause a significant risk to how the system operates.

For example, many applications fail because they are unable to bear the load of a sudden increase in the number of users. In one of his books, *Jakob Nielsen* writes about the user experience that *1 second is about the limit for the user's flow of thought to stay uninterrupted*. Imagine that our system, with its growing load, starts to exceed that limit. Users may stop using the service just because of its performance. Taking this into consideration, non-functional testing is just as important as functional testing.

To cut a long story short, we should always take the following steps for non-functional testing:

1. Decide which non-functional aspects are crucial to our business.
2. For each of them, we must do the following:

    - Specify the tests the same way we did for acceptance testing

    - Add a stage to the CD pipeline (after acceptance testing, while the application is still deployed on the staging environment)

3. The application comes to the release stage only after all the non-functional tests have passed.

Irrespective of the type of non-functional test, the idea is always the same. The approach, however, may differ slightly. Let's examine different test types and the challenges they pose.

Types of non-functional test

Functional tests are always related to the same aspect – the behavior of the system. In contrast, non-functional tests are concerned with a lot of different aspects. Let's discuss the most common system properties and how they can be tested inside the CD process.

Performance testing

Performance tests are the most widely used non-functional tests. They measure the responsiveness and stability of the system. The simplest performance test we can create is one that sends a request to the web service and measures its **round-trip time (RTT)**.

There are different definitions of performance testing. They are often meant to include load, stress, and scalability testing. Sometimes, they are also described as white-box tests. In this book, we will define performance testing as the most basic form of black-box test to measure the latency of the system.

For performance testing, we can use a dedicated framework (for Java, the most popular is JMeter) or just use the same tool we used for our acceptance tests. A simple performance test is usually added as a pipeline stage, just after the acceptance tests. Such a test should fail if the RTT exceeds the given limit and it detects bugs that slow down our service.

> Tip
>
> The JMeter plugin for Jenkins can show performance trends over time.

Load testing

Load tests are used to check how the system functions when there are a lot of concurrent requests. While a system can be very fast with a single request, this doesn't mean that it works fast enough with 1,000 requests being worked on at the same time. During load testing, we measure the average request-response time of many concurrent calls, which are usually performed from many machines. Load testing is a very common QA phase in the release cycle. To automate it, we can use the same tools that we do when conducting a simple performance test; however, in the case of larger systems, we may need a separate client environment to perform a large number of concurrent requests.

Stress testing

Stress testing, also called **capacity testing** or **throughput testing**, is a test that determines how many concurrent users can access our service. It may sound the same as load testing, but in the case of load testing, we set the number of concurrent users (throughput) to a given number, check the response time (latency), and make the build fail if that limit is exceeded. During stress testing, however, we keep the latency constant and increase the throughput to discover the maximum number of concurrent calls when the system is still operable. Therefore, the result of a stress test may be a notification that our system can handle 10,000 concurrent users, which helps us prepare for the peak usage time.

Stress testing is not well suited for the CD process because it requires long tests with an increasing number of concurrent requests. It should be prepared as a separate script of a separate Jenkins pipeline and triggered on demand when we know that the code change can cause performance issues.

Scalability testing

Scalability testing explains how latency and throughput change when we add more servers or services. The perfect characteristic would be linear, which means that if we have one server and the average request-response time is 500 ms when it's used by 100 parallel users, then adding another server would keep the response time the same and allow us to add another 100 parallel users. In reality, it's often hard to achieve this because of the need to keep data consistent between servers.

Scalability testing should be automated and provide a graph that shows the relationship between the number of machines and the number of concurrent users. Such data helps determine the limits of the system and the point at which adding more machines doesn't help.

Scalability tests, similar to stress tests, are hard to put into the CD pipeline and should be kept separate.

Soak testing

Soak tests, also called **endurance tests** or **longevity tests**, run the system for a long time to see if the performance drops after a certain period. They detect memory leaks and stability issues. Since they require a system to run for a long time, it doesn't make sense to run them inside the CD pipeline.

Security testing

Security testing deals with different aspects related to security mechanisms and data protection. Some security aspects are purely functional requirements, such as authentication, authorization, and role assignment. These elements should be checked the same way as any other functional requirement – during the acceptance test phase. Other security aspects are non-functional; for example, the system should be protected against SQL injection. No client would probably specify such a requirement, but it's implicit.

Security tests should be included in the CD process as a pipeline stage. They can be written using the same frameworks as the acceptance tests or with dedicated security testing frameworks – for example, **behavior-driven development (BDD)** security.

> **Information**
>
> Security should also always be a part of the explanatory testing process, in which testers and security experts detect security holes and add new testing scenarios.

Maintainability testing

Maintainability tests explain how simple a system is to maintain. In other words, they judge code quality. We have already described stages in the commit phase that check test coverage and perform static code analysis. The Sonar tool can also provide an overview of the code quality and the technical debt.

Recovery testing

Recovery testing is a technique that's used to determine how quickly the system can recover after it's crashed because of a software or hardware failure. The best case would be if the system doesn't fail at all, even if a part of its service is down. Some companies even perform production failures on purpose to check if they can survive a disaster. The most well-known example is Netflix and their Chaos Monkey tool, which randomly terminates instances of the production environment. Such an approach forces engineers to write code that makes systems resilient to failures.

Recovery testing is not part of the CD process, but rather a periodic event that checks its overall health.

> **Tip**
>
> You can read more about Chaos Monkey at `https://github.com/Netflix/chaosmonkey`.

Many more nonfunctional test types are closer to or further from the code and the CD process. Some of them relate to the law, such as compliance testing, while others are related to documentation or internationalization. There's also usability testing and volume testing (which check whether the system behaves well when it's handling large amounts of data). Most of these tests, however, have no part in the CD process.

Non-functional challenges

Non-functional aspects pose new challenges to software development and delivery. Let's go over some of them now:

- **Long test runs**: The tests can take a long time to run and may need a special execution environment.

- **Incremental nature**: It's hard to set the limit value when the test should fail (unless the SLA is well-defined). Even if the edge limit is set, the application would probably incrementally approach the limit. In most cases, no code changes will cause the test to fail.

- **Vague requirements**: Users usually don't have much input when it comes to non-functional requirements. They may provide some guidelines concerning the request-response time or the number of users; however, they probably won't know much about maintainability, security, or scalability.

- **Multiplicity**: There are a lot of different non-functional tests and choosing which should be implemented means making some compromises.

The best approach to address non-functional aspects is to perform the following steps:

1. Make a list of all the non-functional test types.

2. Explicitly cross out the tests you don't need for your system. There may be a lot of reasons you don't need one kind of test, such as the following:

   - The service is super small, and a simple performance test is enough.

   - The system is internal only and exclusively available for read-only purposes, so it may not need any security checks.

   - The system is designed for one machine only and does not need any scaling.

   - The cost of creating certain tests is too high.

3. Split your tests into two groups:

   - **Continuous Delivery**: It is possible to add it to the pipeline.

   - **Analysis**: It is not possible to add it to the pipeline because of its execution time, nature, or associated cost.

4. For the CD group, implement the related pipeline stages.

5. For the analysis group, do the following:

- Create automated tests

- Schedule when they should be run

- Schedule meetings to discuss their results and take action

> **Tip**
>
> A very good approach is to have a nightly build with the long tests that don't fit the CD pipeline. Then, it's possible to schedule a weekly meeting to monitor and analyze the trends of system performance.

As we can see, there are many types of non-functional tests, and they pose additional challenges to the delivery process. Nevertheless, for the sake of the stability of our system, these tests should never be skipped. The technical implementation differs based on the test's type, but in most cases, they can be implemented similarly to functional acceptance tests and should be run against the staging environment.

> **Tip**
>
> If you're interested in the topic of non-functional testing, system properties, and system stability, then read the book *Release It!*, by *Michael T. Nygard*.

Now that we've discussed the nonfunctional testing, let's look at another aspect that we haven't looked at in too much detail – application versioning.

Application versioning

So far, throughout every Jenkins build, we have created a new Docker image, pushed it into the Docker registry, and used the *latest* version throughout the process. However, such a solution has at least three disadvantages:

- If, during the Jenkins build, after the acceptance tests, someone pushes a new version of the image, then we can end up releasing the untested version.

- We always push an image that's named in the same way so that, effectively, it is overwritten in the Docker registry.

- It's very hard to manage images without versions just by using their hashed-style IDs.

What is the recommended way of managing Docker image versions alongside the CD process? In this section, we'll look at the different versioning strategies and learn how to create versions in the Jenkins pipeline.

Versioning strategies

There are different ways to version applications.

Let's discuss the most popular solutions that can be applied alongside the CD process (when each commit creates a new version):

- **Semantic versioning**: The most popular solution is to use sequence-based identifiers (usually in the form of $x.y.z$). This method requires a commit to be made to the repository by Jenkins to increase the current version number, which is usually stored in the build file. This solution is well supported by Maven, Gradle, and other build tools. The identifier usually consists of three numbers:

  - x: This is the major version; the software does not need to be backward compatible when this version is incremented.

  - y: This is the minor version; the software needs to be backward compatible when the version is incremented.

  - z: This is the build number (also called the **patch version**); this is sometimes also considered as a backward-and forward-compatible change.

- **Timestamp**: Using the date and time of the build for the application version is less verbose than sequential numbers, but it's very convenient in the case of the CD process because it does not require Jenkins to commit it back to the repository.

- **Hash**: A randomly generated hash version shares the benefit of the date-time and is probably the simplest solution possible. The drawback is that it's not possible to look at two versions and tell which is the latest one.

- **Mixed**: There are many variations of the solutions described earlier – for example, the major and minor versions with the date-time.

All of these solutions can be used alongside the CD process. Semantic versioning, however, requires a commit to be made to the repository from the build execution so that the version is increased in the source code repository.

> **Information**
>
> Maven (and other build tools) popularized version snapshotting, which added a SNAPSHOT suffix to the versions that haven't been released and have been kept just for the development process. Since CD means releasing every change, there are no snapshots.

Now, let's learn how to adapt versioning in the Jenkins pipeline.

Versioning in the Jenkins pipeline

As we mentioned earlier, there are different possibilities when it comes to using software versioning, and each of them can be implemented in Jenkins.

As an example, let's use the date-time.

> **Information**
>
> To use the timestamp information from Jenkins, you need to install the Build Timestamp plugin and set the timestamp format in the Jenkins configuration under **Manage Jenkins** | **Configure System** | **Build Timestamp**. You can set the pattern to, for example, yyyyMMdd-HHmm.

Everywhere we use the Docker image, we need to add the ${BUILD_TIMESTAMP} tag suffix.

For example, the Docker build stage should look like this:

```
sh "docker build -t leszko/calculator:${BUILD_TIMESTAMP} ."
```

After making these changes, when we run the Jenkins build, the image should be tagged with the timestamp's version in our Docker registry.

With versioning completed, we are finally ready to complete the CD pipeline.

Completing the continuous delivery pipeline

Now that we've covered Ansible, environments, non-functional testing, and versioning, we are ready to extend the Jenkins pipeline and finalize a simple, but complete, CD pipeline.

Follow these steps:

1. Create the inventory of staging and production environments.
2. Use version in the Kubernetes deployment.

3. Use a remote Kubernetes cluster as the staging environment.

4. Update the acceptance tests so that they use the staging Kubernetes cluster.

5. Release the application to the production environment.

6. Add a smoke test that makes sure the application was released successfully.

Let's start by creating an inventory.

Inventory

We looked at the inventory file in the previous chapter while describing Ansible. To generalize this concept, an inventory contains a list of environments that describe how to access them. In this example, we'll use Kubernetes directly, so the Kubernetes configuration file, which is usually stored in .kube/config, will act as the inventory.

> **Information**
>
> As we explained in the previous chapter, depending on your needs, you may use kubectl directly or via Ansible or Terraform. These approaches are suitable for the CD pipeline.

Let's configure two Kubernetes clusters – staging and production. Your .kube/ config file should look similar to the following one:

```
apiVersion: v1
clusters:
- cluster:
    certificate-authority-data: LS0tLS1CR...
    server: https://35.238.191.252
  name: staging
- cluster:
    certificate-authority-data: LS0tLS1CR...
    server: https://35.232.61.210
  name: production
contexts:
- context:
    cluster: staging
    user: staging
  name: staging
- context:
```

```
      cluster: production
      user: production
    name: production
users:
- name: staging
  user:
      token: eyJhbGciOiJSUzI1NiIsImtpZCI6I...
- name: production
  user:
      token: eyJ0eXAiOiJKV1QiLCJhbGciOiJSU...
```

The Kubernetes configuration stores the following information for each cluster:

- `cluster`: The address of the cluster (Kubernetes master endpoint) and its CA certificate

- `context`: The binding of the cluster and user

- `user`: The authorization data to access the Kubernetes cluster

> **Tip**
>
> The simplest way to create two Kubernetes clusters is to use **Google Kubernetes Engine (GKE)**, then configure `kubectl` using `gcloud container clusters get-credentials`, and finally rename the cluster context with `kubectl config rename-context <original-context-name> staging`. Note that you may also need to create a GCP Firewall rule to allow traffic into your Kubernetes nodes.

You also need to make sure that the Kubernetes configuration is available on the Jenkins agent nodes. As we mentioned in the previous sections, think carefully about your security so that no unauthorized persons can access your environments via the Jenkins agent.

Now that we've defined the inventory, we can prepare the Kubernetes deployment configuration so that it can work with application versioning.

Versioning

Kubernetes YAML files are the same as what we defined in the previous chapters. The only difference is that we need to introduce a template variable for the application version. Let's make one change in the `deployment.yaml` file:

```
image: leszko/calculator:{{VERSION}}
```

Then, we can fill the version in `Jenkinsfile`:

```
stage("Update version") {
    steps {
        sh "sed -i 's/{{VERSION}}/${BUILD_TIMESTAMP}/g'
deployment.yaml"
    }
}
```

Now, we can change acceptance testing to use the remote staging environment.

The remote staging environment

Depending on our needs, we could test the application by running it on the local Docker host (as we did previously) or using the remote (and clustered) staging environment. The former solution is closer to what happens in production, so it can be considered a better one.

To do this, we need to change the command we use from `docker` to `kubectl`. Let's modify the related part of our `Jenkinsfile`:

```
stage("Deploy to staging") {
    steps {
        sh "kubectl config use-context staging"
        sh "kubectl apply -f hazelcast.yaml"
        sh "kubectl apply -f deployment.yaml"
        sh "kubectl apply -f service.yaml"
    }
}
```

First, we switched `kubectl` to use the `staging` context. Then, we deployed the Hazelcast server. Finally, we deployed `Calculator` into the Kubernetes server. At this point, we have a fully functional application in our staging environment. Let's see how we need to modify the acceptance testing stage.

The acceptance testing environment

The `Acceptance test` stage looks the same as it did in the previous chapter. The only thing we need to change is the IP and port of our service to the one from the remote Kubernetes cluster. As we explained in *Chapter 6, Clustering with Kubernetes*, the way you should do this depends on your Kubernetes Service type. We used `NodePort`, so we need to make the following change in `Jenkinsfile`:

```
stage("Acceptance test") {
    steps {
        sleep 60
        sh "chmod +x acceptance-test.sh && ./acceptance-test.
sh"
    }
}
```

The `acceptance-test.sh` script should look as follows:

```
#!/bin/bash
set -x

NODE_IP=$(kubectl get nodes -o jsonpath='{ $.items[0].status.
addresses[?
        (@.type=="ExternalIP")].address }')
NODE_PORT=$(kubectl get svc calculator-service -o=jsonpath='{.
spec.ports[0].nodePort}')
./gradlew acceptanceTest -Dcalculator.url=http://${NODE_
IP}:${NODE_PORT}
```

First, we used `sleep` to wait for our application to be deployed. Then, using `kubectl`, we fetched the IP address (`NODE_IP`) and the port (`NODE_PORT`) of our service. Finally, we executed the acceptance testing suite.

> **Tip**
> If you use Minishift for your Kubernetes cluster, then you can fetch NODE_
> IP using `minishift ip`. If you use Docker for Desktop, then your IP will
> be `localhost`.

Now that all our tests are in place, it's time to release the application.

Release

The production environment should be as close to the staging environment as possible. The Jenkins stage for the release should also be as close as possible to the Deploy to staging step.

In the simplest scenario, the only difference will be the Kubernetes configuration context and the application configuration (for example, in the case of a Spring Boot application, we would set a different Spring profile, which results in taking a different application. properties file). In our case, there are no application properties, so the only difference is the kubectl context:

```
stage("Release") {
    steps {
        sh "kubectl config use-context production"
        sh "kubectl apply -f hazelcast.yaml"
        sh "kubectl apply -f deployment.yaml"
        sh "kubectl apply -f service.yaml"
    }
}
```

Once the release has been done, we may think that everything is complete; however, one stage is missing – smoke testing.

Smoke testing

A smoke test is a very small subset of acceptance tests whose only purpose is to check that the release process is completed successfully; otherwise, we could have a situation where the application is perfectly fine, but where there is an issue in the release process, so we may end up with a non-working production environment.

The smoke test is usually defined in the same way as the acceptance test. So, the Smoke test stage in the pipeline should look like this:

```
stage("Smoke test") {
    steps {
        sleep 60
        sh "chmod +x smoke-test.sh && ./smoke-test.sh"
    }
}
```

Once everything has been set up, the CD build should run automatically, and the application should be released to production. With that, we have finished analyzing the CD pipeline in its simplest, but fully productive, form.

Complete Jenkinsfile

To summarize, in the past few chapters, we have gone through quite a few stages that have resulted in us creating a complete CD pipeline that can be used in many projects.

The following is the complete `Jenkinsfile` for the `Calculator` project:

```
pipeline {
  agent any

  triggers {
    pollSCM('* * * * *')
  }

  stages {
    stage("Compile") { steps { sh "./gradlew compileJava" } }
    stage("Unit test") { steps { sh "./gradlew test" } }

    stage("Code coverage") { steps {
      sh "./gradlew jacocoTestReport"
      sh "./gradlew jacocoTestCoverageVerification"
    } }

    stage("Static code analysis") { steps {
      sh "./gradlew checkstyleMain"
    } }

    stage("Build") { steps { sh "./gradlew build" } }

    stage("Docker build") { steps {
      sh "docker build -t leszko/calculator:${BUILD_TIMESTAMP}
."
    } }
```

```groovy
    stage("Docker push") { steps {
      sh "docker push leszko/calculator:${BUILD_TIMESTAMP}"
    } }

    stage("Update version") { steps {
        sh "sed -i 's/{{VERSION}}/${BUILD_TIMESTAMP}/g'
deployment.yaml"
    } }

    stage("Deploy to staging") { steps {
      sh "kubectl config use-context staging"
      sh "kubectl apply -f hazelcast.yaml"
      sh "kubectl apply -f deployment.yaml"
      sh "kubectl apply -f service.yaml"
    } }

    stage("Acceptance test") { steps {
      sleep 60
      sh "chmod +x acceptance-test.sh && ./acceptance-test.sh"
    } }

    // Performance test stages

    stage("Release") { steps {
      sh "kubectl config use-context production"
      sh "kubectl apply -f hazelcast.yaml"
      sh "kubectl apply -f deployment.yaml"
      sh "kubectl apply -f service.yaml"
    } }

    stage("Smoke test") { steps {
      sleep 60
      sh "chmod +x smoke-test.sh && ./smoke-test.sh"
    } }
  }
}
```

The preceding code is a declarative description of the whole CD process, which starts with checking out the code and ends with releasing it to production. Congratulations – with this code, you have completed the main goal of this book, which is to create a CD pipeline!

Summary

In this chapter, we completed the CD pipeline, which means we can finally release the application. The following are the key takeaways from this chapter:

- When it comes to CD, two environments are indispensable: staging and production.
- Non-functional tests are an essential part of the CD process and should always be considered as pipeline stages.
- Non-functional tests that don't fit the CD process should be used as periodic tasks to monitor the overall performance trends.
- Applications should always be versioned; however, the versioning strategy depends on the type of application.
- A minimal CD pipeline can be implemented as a sequence of scripts that ends with two stages: release and smoke test.
- The smoke test should always be added as the last stage of the CD pipeline to check whether the release was successful.

In the next chapter, we will look at some of the advanced aspects of the CD pipeline.

Exercises

In this chapter, we have covered a lot of new aspects of the CD pipeline. To help you understand these concepts, we recommend that you complete the following exercises:

1. Add a performance test that tests the `hello world` service:

    I.   The `hello world` service can be taken from the previous chapter.

    II. Create a `performance-test.sh` script that makes 100 calls and checks whether the average request-response time is less than 1 second.

    III. You can use Cucumber or the `curl` command for the script.

2. Create a Jenkins pipeline that builds the `hello world` web service as a versioned Docker image and performs performance tests:

    I.   Create a `Docker build` (and `Docker push`) stage that builds the Docker image with the `hello world` service and adds a timestamp as a version tag.

    II.   Use the Kubernetes deployment from the previous chapters to deploy the application.

    III. Add the `Deploy to staging` stage, which deploys the image to the remote machine.

    IV. Add the `Performance testing` stage, which executes `performance-test.sh`.

    V.   Run the pipeline and observe the results.

Questions

To check your knowledge of this chapter, answer the following questions:

1. Name at least three different types of software environments.
2. What is the difference between the staging and QA environments?
3. Name at least five types of non-functional tests.
4. Should all non-functional tests be part of the CD pipeline?
5. Name at least two types of application versioning strategies.
6. What is a smoke test?

Further reading

To learn more about the CD pipeline, please refer to the following resources:

- *Sameer Paradkar: Mastering Non-Functional Requirements*: `https://www.packtpub.com/application-development/mastering-non-functional-requirements`.

- *Sander Rossel: Continuous Integration, Delivery, and Deployment*: `https://www.packtpub.com/application-development/continuous-integration-delivery-and-deployment`.

9
Advanced Continuous Delivery

In the previous chapters, we started with nothing and ended with a complete continuous delivery pipeline. Now, it's time to present a mixture of different aspects that are also very important in the continuous delivery process, but which haven't been described yet.

This chapter covers the following points:

- Managing database changes
- Pipeline patterns
- Release patterns
- Working with legacy systems

Technical requirements

To follow along with the instructions in this chapter, you'll need the following:

- Java 8+
- A Jenkins instance

All the examples and solutions to the exercises can be found on GitHub at `https://github.com/PacktPublishing/Continuous-Delivery-With-Docker-and-Jenkins-3rd-Edition/tree/main/Chapter09`.

Code in Action videos for this chapter can be viewed at `https://bit.ly/3NVVOyi`.

Managing database changes

So far, we have focused on a continuous delivery process that was applied to a web service. A simple factor in this was that web services are inherently stateless. This fact means that they can easily be updated, restarted, cloned in many instances, and recreated from the given source code. A web service, however, is usually linked to its stateful part: a database that poses new challenges to the delivery process. These challenges can be grouped into the following categories:

- **Compatibility**: The database schema, and the data itself, must be compatible with the web service all the time.

- **Zero-downtime deployment**: In order to achieve zero-downtime deployment, we use rolling updates, which means that a database must be compatible with two different web service versions at the same time.

- **Rollback**: A rollback of a database can be difficult, limited, or sometimes even impossible because not all operations are reversible (for example, removing a column that contains data).

- **Test data**: Database-related changes are difficult to test because we need test data that is very similar to production data.

In this section, I will explain how to address these challenges so that the continuous delivery process will be as safe as possible.

Understanding schema updates

If you think about the delivery process, it's not really the data itself that causes difficulties, because we don't usually change the data when we deploy an application. The data is something that is collected while the system is live in production, whereas, during deployment, we only change the way we store and interpret this data. In other words, in the context of the continuous delivery process, we are interested in the structure of the database, not exactly in its content. This is why this section concerns mainly relational databases (and their schemas) and focuses less on other types of storage, such as NoSQL databases, where there is no structure definition.

To better understand this, think of Hazelcast, which we have already used in this book. It stored the cached data so, effectively, it was a database. Nevertheless, it required zero effort from the continuous delivery perspective since it didn't have any data structure. All it stored were the key-value entries, which do not evolve over time.

> **Information**
>
> NoSQL databases usually don't have any restricting schema, and therefore, they simplify the continuous delivery process because there is no additional schema update step required. This is a huge benefit; however, it doesn't necessarily mean that writing applications with NoSQL databases is simpler because we have to put more effort into data validation in the source code.

Relational databases have static schemas. If we would like to change it (for example, to add a new column to the table), we need to write and execute a SQL **data definition language** (DDL) script. Doing this manually for every change requires a lot of work and leads to error-prone solutions in which the operations team has to keep the code and the database structure in sync. A much better solution is to automatically update the schema in an incremental manner. Such a solution is called **database migration**.

Introducing database migration

Database schema migration is a process of incremental changes to the relational database structure. Let's take a look at the following diagram to understand it better:

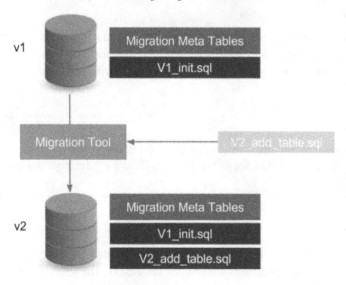

Figure 9.1 – Database schema migration

The database in version *v1* has the schema defined by the `V1_init.sql` file. Additionally, it stores the metadata related to the migration process, for example, its current schema version and the migration changelog. When we want to update the schema, we provide the changes in the form of a SQL file, such as `V2_add_table.sql`. Then, we need to run the migration tool that executes the given SQL file on the database (it also updates the metatables). In effect, the database schema is a result of all subsequently executed SQL migration scripts. Next, we will see an example of migration.

> **Information**
>
> Migration scripts should be stored in the version control system, usually in the same repository as the source code.

Migration tools and the strategies they use can be divided into two categories:

- **Upgrade and downgrade**: This approach (as used by the Ruby on Rails framework, for example) means that we can migrate up (from *v1* to *v2*) and down (from *v2* to *v1*). It allows the database schema to roll back, which may sometimes end in data loss (if the migration is logically irreversible).

- **Upgrade only**: This approach (as used by the Flyway tool, for example) only allows us to migrate up (from *v1* to *v2*). In many cases, the database updates are not reversible, for example, when removing a table from the database. Such a change cannot be rolled back, because even if we recreate the table, we have already lost all the data.

There are many database migration tools available on the market, the most popular of which are **Flyway**, **Liquibase**, and **Rail Migrations** (from the Ruby on Rails framework). As a next step to understanding how such tools work, we will look at an example based on the Flyway tool.

> **Information**
>
> There are also commercial solutions provided for the particular databases, for example, Redgate (for SQL Server) and Optim Database Administrator (for DB2).

Using Flyway

Let's use Flyway to create a database schema for the calculator web service. The database will store the history of all operations that were executed on the service: the first parameter, the second parameter, and the result.

We show how to use the SQL database and Flyway in three steps:

1. Configuring the Flyway tool to work with Gradle
2. Defining the SQL migration script to create the calculation history table
3. Using the SQL database inside the Spring Boot application code

Let's get started.

Configuring Flyway

In order to use Flyway with Gradle, we need to add the following content to the `build.gradle` file:

```
buildscript {
    dependencies {
        classpath('com.h2database:h2:1.4.200')
    }
}
. . .
plugins {
    id "org.flywaydb.flyway" version "8.5.0"
}
. . .
flyway {
    url = 'jdbc:h2:file:/tmp/calculator'
    user = 'sa'
}
```

Here are some quick comments on the configuration:

- We used the H2 database, which is an in-memory (and file-based) database.
- We store the database in the `/tmp/calculator` file.
- The default database user is called `sa` (system administrator).

> **Tip**
> In the case of other SQL databases (for example, MySQL), the configuration would be very similar. The only difference is in the Gradle dependencies and the JDBC connection.

After this configuration is applied, we should be able to run the Flyway tool by executing the following command:

```
$ ./gradlew flywayMigrate -i
```

The command created the database in the /tmp/calculator.mv.db file. Obviously, it has no schema, since we haven't defined anything yet.

> **Information**
>
> Flyway can be used as a command-line tool, via the Java API, or as a plugin for the popular building tools Gradle, Maven, and Ant.

Defining the SQL migration script

The next step is to define the SQL file that adds the calculation table to the database schema. Let's create the src/main/resources/db/migration/V1__Create_calculation_table.sql file, with the following content:

```
create table CALCULATION (
    ID      int not null auto_increment,
    A       varchar(100),
    B       varchar(100),
    RESULT  varchar(100),
    primary key (ID)
);
```

Note the migration file naming convention, <version>__<change_description>.sql. The SQL file creates a table with four columns, ID, A, B, and RESULT. The ID column is an automatically incremented primary key of the table. Now, we are ready to run the flyway command to apply the migration:

```
$ ./gradlew flywayMigrate -i
...
Migrating schema "PUBLIC" to version "1 - Create calculation table"
Successfully applied 1 migration to schema "PUBLIC", now at version v1 (execution time 00:00.018s)
```

The command automatically detected the migration file and executed it on the database.

> **Information**
>
> The migration files should be always kept in the version control system, usually with the source code.

Accessing the database

We have executed our first migration, so the database is prepared. To see the complete example, we should also adapt our project so that it can access the database.

Let's first configure the Gradle dependencies to use `h2database` from the Spring Boot project:

1. We can do this by adding the following lines to the `build.gradle` file:

```
dependencies {
    implementation 'org.springframework.boot:spring-boot-
starter-data-jpa'
    implementation 'com.h2database:h2:1.4.200'
}
```

2. The next step is to set up the database location and the startup behavior in the `src/main/resources/application.properties` file:

```
spring.datasource.url=jdbc:h2:file:/tmp/calculator;DB_
CLOSE_ON_EXIT=FALSE
spring.jpa.hibernate.ddl-auto=validate
spring.datasource.username=sa
```

The second line means that Spring Boot will not try to automatically generate the database schema from the source code model. On the contrary, it will only validate if the database schema is consistent with the Java model.

3. Now, let's create the Java ORM entity model for the calculation in the new `src/main/java/com/leszko/calculator/Calculation.java` file:

```java
package com.leszko.calculator;
import javax.persistence.Entity;
import javax.persistence.GeneratedValue;
import javax.persistence.GenerationType;
import javax.persistence.Id;


@Entity
public class Calculation {
```

```
    @Id
    @GeneratedValue(strategy= GenerationType.IDENTITY)
    private Integer id;
    private String a;
    private String b;
    private String result;


    protected Calculation() {}


    public Calculation(String a, String b, String result)
{
        this.a = a;
        this.b = b;
        this.result = result;
    }
}
```

The Entity class represents the database mapping in the Java code. A table is expressed as a class, with each column as a field. The next step is to create the repository for loading and storing the Calculation entities.

4. Let's create the src/main/java/com/leszko/calculator/ CalculationRepository.java file:

```
package com.leszko.calculator;
import org.springframework.data.repository.
CrudRepository;


public interface CalculationRepository extends
CrudRepository<Calculation, Integer> {}
```

5. Finally, we can use the Calculation and CalculationRepository classes to store the calculation history. Let's add the following code to the src/main/java/ com/leszko/calculator/CalculatorController.java file:

```
...
class CalculatorController {

    ...


    @Autowired
```

```
    private CalculationRepository calculationRepository;

    @RequestMapping("/sum")
    String sum(@RequestParam("a") Integer a, @
RequestParam("b") Integer b) {
        String result = String.valueOf(calculator.sum(a,
b));
        calculationRepository.save(new Calculation(a.
toString(), b.toString(), result));
        return result;
    }
}
```

6. Now, we can finally start the service, for example, using the following command:

```
$ ./gradlew bootRun
```

When we have started the service, we can send a request to the /sum endpoint. As a result, each summing operation is logged into the database.

> **Tip**
>
> If you would like to browse the database content, you can add spring.
> h2.console.enabled=true to the application.properties
> file, and then browse the database via the /h2-console endpoint.

We explained how the database schema migration works and how to use it inside a Spring Boot project built with Gradle. Now, let's take a look at how it integrates within the continuous delivery process.

Changing the database in continuous delivery

The first approach to use database updates inside the continuous delivery pipeline is to add a stage within the migration command execution. This simple solution works correctly for many cases; however, it has two significant drawbacks:

*   **Rollback**: As mentioned before, it's not always possible to roll back the database change (Flyway doesn't support downgrades at all). Therefore, in the case of service rollback, the database becomes incompatible.

*   **Downtime**: The service update and the database update are not executed at exactly the same time, which causes downtime.

This leads us to two constraints that we will need to address:

- The database version needs to be compatible with the service version all the time.

- The database schema migration is not reversible.

We will address these constraints for two different cases: backward-compatible updates and non-backward-compatible updates.

Backward-compatible changes

Backward-compatible changes are simpler. Let's look at the following figure to see how they work:

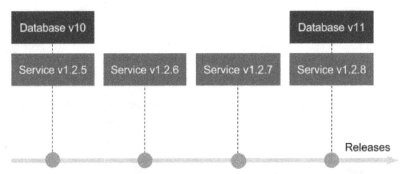

Figure 9.2 – Backward-compatible database migration

Suppose that the **Database v10** schema migration is backward-compatible. If we need to roll back the **Service v1.2.8** release, then we deploy **Service v1.2.7**, and there is no need to do anything with the database (database migrations are not reversible, so we keep **Database v11**). Since the schema update is backward-compatible, **Service v.1.2.7** works perfectly fine with **Database v11**. The same applies if we need to roll back to **Service v1.2.6**, and so on. Now, suppose that **Database v10** and all other migrations are backward-compatible; then we could roll back to any service version, and everything would work correctly.

There is also no problem with downtime. If the database migration has zero-downtime itself, then we can execute it first, and then use the rolling updates for the service.

Let's look at an example of a backward-compatible change. We will create a schema update that adds a `created_at` column to the calculation table. The `src/main/resources/db/migration/V2__Add_created_at_column.sql` migration file looks as follows:

```
alter table CALCULATION
add CREATED_AT timestamp;
```

Aside from the migration script, the calculator service requires a new field in the `Calculation` class:

```
...
private Timestamp createdAt;
...
```

We also need to adjust its constructor, and then its usage in the `CalculatorController` class:

```
calculationRepository.save(new Calculation(a.toString(),
b.toString(), result, Timestamp.from(Instant.now())));
```

After running the service, the calculation history is stored with the `created_at` column. Note that the change is backward-compatible because, even if we reverted the Java code and left the `created_at` column in the database, everything would work perfectly fine (the reverted code does not address the new column at all).

Non-backward-compatible changes

Non-backward-compatible changes are way more difficult. Looking at the previous diagram, if the **v11** database change was backward-incompatible, it would be impossible to roll back the service to **1.2.7**. In this case, how can we approach non-backward-compatible database migrations so that rollbacks and zero-downtime deployments would be possible?

To cut a long story short, we can address this issue by converting a non-backward-compatible change into a change that is backward-compatible for a certain period of time. In other words, we need to put in the extra effort and split the schema migration into two parts:

- Backward-compatible update executed now, which usually means keeping some redundant data
- Non-backward-compatible update executed after the rollback period time that defines how far back we can revert our code

To better illustrate this, let's look at the following diagram:

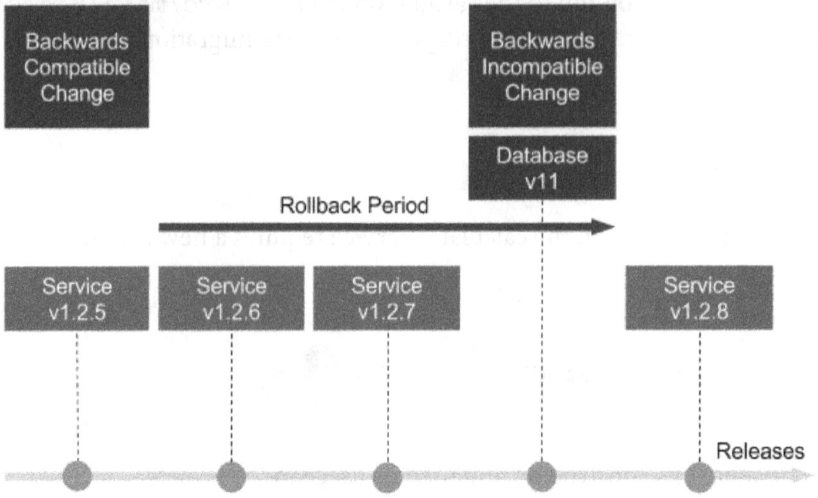

Figure 9.3 – Non-backward-compatible database migration

Let's consider an example of dropping a column. A proposed method would include two steps:

- Stop using the column in the source code (**v1.2.5**, backward-compatible update, executed first).

- Drop the column from the database (**v11**, non-backward-compatible update, executed after the rollback period).

All service versions until **Database v11** can be rolled back to any previous version; the services starting from **Service v1.2.8** can only be rolled back within the rollback period. Such an approach may sound trivial because all we did was delay the column removal from the database. However, it addresses both the rollback issue and the zero-downtime deployment issue. As a result, it reduces the risk associated with the release. If we adjust the rollback period to a reasonable amount of time (for example, in the case of multiple releases per day to 2 weeks), then the risk is negligible. We don't usually roll many versions back.

Dropping a column was a very simple example. Let's take a look at a more difficult scenario and rename the result column in our calculator service. We show how to do this in a few steps:

1. Adding a new column to the database
2. Changing the code to use both columns

3. Merging the data in both columns

4. Removing the old column from the code

5. Dropping the old column from the database

Let's look at these steps in detail.

Adding a new column to the database

Let's suppose that we need to rename the `result` column to `sum`. The first step is to add a new column that will be a duplicate. We must create a `src/main/resources/db/ migration/V3__Add_sum_column.sql` migration file:

```
alter table CALCULATION
add SUM varchar(100);
```

As a result, after executing the migration, we will have two columns: `result` and `sum`.

Changing the code to use both columns

The next step is to rename the column in the source code model and to use both database columns for the `set` and `get` operations. We can change it in the `Calculation` class:

```
public class Calculation {
    ...
    private String sum;
    ...
    public Calculation(String a, String b, String sum,
Timestamp createdAt) {
        this.a = a;
        this.b = b;
        this.sum = sum;
        this.result = sum;
        this.createdAt = createdAt;
    }

    public String getSum() {
        return sum != null ? sum : result;
    }
}
```

> **Tip**
>
> To be 100% accurate, in the `getSum()` method, we should compare
> something like the last modification column date. (It's not exactly necessary to
> always take the new column first.)

From now on, every time we add a row into the database, the same value is written to both
the `result` and `sum` columns. While reading `sum`, we first check whether it exists in the
new column, and if not, we read it from the old column.

> **Tip**
>
> The same result can be achieved with the use of database triggers that would
> automatically write the same values into both columns.

All the changes that we have made so far are backward-compatible, so we can roll back the
service anytime we want, to any version we want.

Merging the data in both columns

This step is usually done after some time when the release is stable. We need to copy the
data from the old `result` column into the new `sum` column. Let's create a migration file
called `V4__Copy_result_into_sum_column.sql`:

```
update CALCULATION
set CALCULATION.sum = CALCULATION.result
where CALCULATION.sum is null;
```

We still have no limits for the rollback; however, if we need to deploy the version before
the change in *step 2*, then this database migration needs to be repeated.

Removing the old column from the code

At this point, we already have all data in the new column, so we can start to use it without
the old column in the data model. In order to do this, we need to remove all code related
to `result` in the `Calculation` class so that it would look as follows:

```
public class Calculation {
    ...
    private String sum;
    ...
    public Calculation(String a, String b, String sum,
Timestamp createdAt) {
```

```
        this.a = a;
        this.b = b;
        this.sum = sum;
        this.createdAt = createdAt;
    }

    public String getSum() {
        return sum;
    }
}
}
```

After this operation, we will no longer use the `result` column in the code. Note that this operation is only backward-compatible up to *step 2*. If we need to roll back to *step 1*, then we could lose the data stored after this step.

Dropping the old column from the database

The last step is to drop the old column from the database. This migration should be performed after the rollback period when we are sure we won't need to roll back before *step 4*.

> **Information**
>
> The rollback period can be very long since we aren't using the column from the database anymore. This task can be treated as a cleanup task, so even though it's non-backward-compatible, there is no associated risk.

Let's add the final migration, `V5__Drop_result_column.sql`:

```
alter table CALCULATION
  drop column RESULT;
```

After this step, we will have finally completed the column renaming procedure. Note that the steps we took complicated the operation a little bit in order to stretch it in time. This reduced the risk of backward-incompatible database changes and allowed for zero-downtime deployments.

Separating database updates from code changes

So far, in all images, we showed that database migrations are run with service releases. In other words, each commit (which implies each release) took both database changes and code changes. However, the recommended approach is to make a clear separation that a commit to the repository is either a database update or a code change. This method is presented in the following diagram:

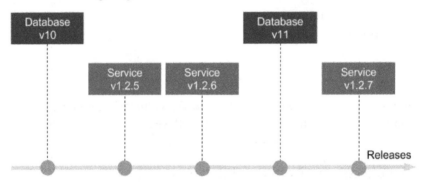

Figure 9.4 – Separating database updates and code changes

The benefit of database-service change separation is that we get the backward-compatibility check for free. Imagine that the **v11** and **v1.2.7** changes concern one logical change, for example, adding a new column to the database. Then, we first commit **Database v11**, so the tests in the continuous delivery pipeline check whether **Database v11** works correctly with **Service v.1.2.6**. In other words, they check whether the **Database v11** update is backward-compatible. Then, we commit the **v1.2.7** change, so the pipeline checks whether **Database v11** works with **Service v1.2.7**.

> **Information**
>
> The database-code separation does not mean that we must have two separate Jenkins pipelines. The pipeline can always execute both, but we should keep it as a good practice that a commit is either a database update or a code change.

To sum up, the database schema changes should never be done manually. Instead, we should always automate them using a migration tool executed as a part of the continuous delivery pipeline. We should also avoid non-backward-compatible database updates, and the best way to ensure this is to commit the database and code changes into the repository separately.

Avoiding a shared database

In many systems, we can spot that the database becomes the central point that is shared between multiple services. In such a case, any update to the database becomes much more challenging, because we need to coordinate it between all services.

For example, imagine we are developing an online shop, and we have a `Customers` table that contains the following columns: `first name`, `last name`, `username`, `password`, `email`, and `discount`. There are three services that are interested in the customer's data:

- **Profile manager**: This enables editing user's data.

- **Checkout processor**: This processes the checkout (reads username and email).

- **Discount manager**: This analyzes the customer's orders and applies a suitable discount.

Let's look at the following diagram that shows this situation:

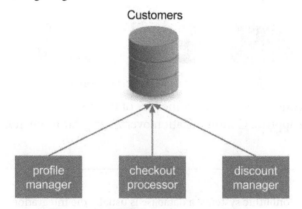

Figure 9.5 – Shared database anti-pattern

The three services are dependent on the same database schema. There are at least two issues with such an approach:

- When we want to update the schema, it must be compatible with all three services. While all backward-compatible changes are fine, any non-backward-compatible update becomes far more difficult, or even impossible.

- Each service has a separate delivery cycle and a separate continuous delivery pipeline. So, *which pipeline should we use for the database schema migrations?* Unfortunately, there is no good answer to this question.

For the reasons mentioned previously, each service should have its own database and the services should communicate via their APIs. Using our example, we could apply the following refactoring:

- The checkout processor should communicate with the profile manager's API to fetch the customer's data.

- The discount column should be extracted to a separate database (or schema), and the discount manager should take ownership.

The refactored version is presented in the following diagram:

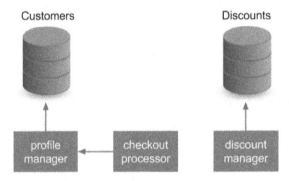

Figure 9.6 – Database per service pattern

Such an approach is consistent with the principles of the microservice architecture and should always be applied. Communication over APIs is far more flexible than direct database access.

> **Information**
>
> In the case of monolithic systems, a database is usually the integration point. Since such an approach causes a lot of issues, it's considered an anti-pattern.

Preparing test data

We have already presented database migrations that keep the database schema consistent between the environments as a side effect. This is because if we run the same migration scripts on the development machine, in the staging environment, or in the production, then we would always get the result in the same schema. However, the data values inside the tables differ. How can we prepare the test data so that it would effectively test our system? This will be the focus of the next section.

The answer to this question depends on the type of test, and it is different for unit testing, integration/acceptance testing, and performance testing. Let's examine each case.

Unit testing

In the case of unit testing, we don't use the real database. We either mock the test data on the level of the persistence mechanism (repositories and data access objects) or we fake the real database with an in-memory database (for example, an H2 database). Since unit tests are created by developers, the exact data values are also usually invented by developers and aren't as important.

Integration/acceptance testing

Integration and acceptance tests usually use the test/staging database, which should be as similar to the production as possible. One approach, adopted by many companies, is to snapshot the production data into staging that guarantees that it is exactly the same. This approach, however, is treated as an anti-pattern, for the following reasons:

- **Test isolation**: Each test operates on the same database, so the result of one test may influence the input of the others.

- **Data security**: Production instances usually store sensitive information and are, therefore, better secured.

- **Reproducibility**: After every snapshot, the test data is different, which may result in flaky tests.

For these reasons, the preferred approach is to manually prepare the test data by selecting a subset of the production data with the customer or the business analyst. When the production database grows, it's worth revisiting its content to see if there are any reasonable cases that should be added.

The best way to add data to the staging database is to use the public API of a service. This approach is consistent with acceptance tests, which are usually black-box. Furthermore, using the API guarantees that the data itself is consistent and simplifies database refactoring by limiting direct database operations.

Performance testing

The test data for the performance testing is usually similar to acceptance testing. One significant difference is the amount of data. In order to test the performance correctly, we need to provide a sufficient volume of input data, as large as is available on the production (during the peak time). For this purpose, we can create data generators, which are usually shared between acceptance and performance tests.

We have covered a lot about databases in the continuous delivery process. Now, let's move to something completely different. Let's move to the topic of improving our Jenkins pipeline using well-known pipeline patterns.

Pipeline patterns

We already know everything necessary to start a project and set up the continuous delivery pipeline with Jenkins, Docker, Kubernetes, Ansible, and Terraform. This section is intended to extend this knowledge with a few of the recommended Jenkins pipeline practices.

Parallelizing pipelines

In this book, we have always executed the pipeline sequentially, stage by stage, step by step. This approach makes it easy to reason the state and the result of the build. If there is first the acceptance test stage and then the release stage, it means that the release won't ever happen until the acceptance tests are successful. Sequential pipelines are simple to understand and usually do not cause any surprises. That's why the first method to solve any problem is to do it sequentially.

However, in some cases, the stages are time-consuming and it's worth running them in parallel. A very good example is performance tests. They usually take a lot of time, so, assuming that they are independent and isolated, it makes sense to run them in parallel. In Jenkins, we can parallelize the pipeline on two different levels:

- **Parallel steps**: Within one stage, parallel processes run on the same agent. This method is simple because all Jenkins workspace-related files are located on one physical machine. However, as always with vertical scaling, the resources are limited to that single machine.

- **Parallel stages**: Each stage can be run in parallel on a separate agent machine that provides horizontal scaling of resources. We need to take care of the file transfer between the environments (using the `stash Jenkinsfile` keyword) if a file created in the previous stage is needed on the other physical machine.

Let's see how this looks in practice. If we want to run two steps in parallel, the `Jenkinsfile` script should look as follows:

```
pipeline {
    agent any
    stages {
        stage('Stage 1') {
```

```
        steps {
            parallel (
                    one: { echo "parallel step 1" },
                    two: { echo "parallel step 2" }
            )
        }
    }
    stage('Stage 2') {
        steps {
            echo "run after both parallel steps are
completed"
        }
    }
  }
}
```

In Stage 1, with the use of the parallel keyword, we execute two parallel steps, one and two. Note that Stage 2 is only executed after both parallel steps are completed. That's why such solutions are perfectly safe to run tests in parallel; we can always be sure that the deployment stage only runs after all parallelized tests have already passed.

The preceding code sample concerned the parallel steps level. The other solution would be to use parallel stages, and therefore, run each stage on a separate agent machine. The decision on which type of parallelism to use usually depends on two factors:

- How powerful the agent machines are
- How much time the given stage takes

As a general recommendation, unit tests are fine to run in parallel steps, but performance tests are usually better off on separate machines.

Reusing pipeline components

When the Jenkinsfile script grows in size and becomes more complex, we may want to reuse its parts between similar pipelines.

For example, we may want to have separate (but similar) pipelines for different environments (development, QA, and production). Another common example in the microservice world is that each service has a very similar `Jenkinsfile`. Then, how do we write `Jenkinsfile` scripts so that we don't repeat the same code all over again? There are two good patterns for this purpose: parameterized builds, and shared libraries. Let's go over them individually.

Build parameters

We already mentioned in *Chapter 4, Continuous Integration Pipeline*, that a pipeline can have input parameters. We can use them to provide different use cases with the same pipeline code. As an example, let's create a pipeline parameterized with the environment type:

```
pipeline {
    agent any

    parameters {
        string(name: 'Environment', defaultValue: 'dev',
    description: 'Which environment (dev, qa, prod)?')
    }

    stages {
        stage('Environment check') {
            steps {
                echo "Current environment: ${params.
    Environment}"
            }
        }
    }
}
```

The build takes one input parameter, `Environment`. Then, all we do in this step is print the parameter. We can also add a condition to execute different code for different environments.

With this configuration, when we start the build we will see a prompt for the input parameter, as follows:

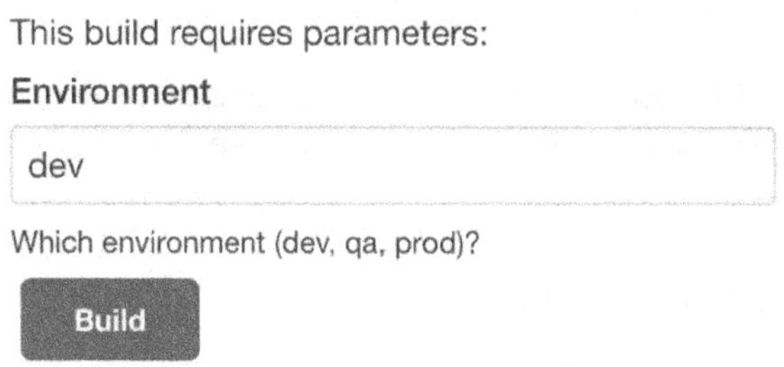

This build requires parameters:

Environment

dev

Which environment (dev, qa, prod)?

Build

Figure 9.7 – Jenkins parametrized build

A parameterized build can help us reuse the pipeline code for scenarios that differ just a little bit. However, this feature should not be overused, because too many conditions can make a `Jenkinsfile` difficult to understand.

Shared libraries

The other solution to reuse the pipeline is to extract its parts into a shared library.

A shared library is a Groovy code that is stored as a separate, source-controlled project. This code can later be used in many `Jenkinsfile` scripts as pipeline steps. To make it clear, let's take a look at an example. A shared library technique always requires three steps:

1. Create a shared library project.
2. Configure the shared library in Jenkins.
3. Use the shared library in a `Jenkinsfile`.

Creating a shared library project

We start by creating a new Git project, in which we put the shared library code. Each Jenkins step is expressed as a Groovy file located in the `vars` directory.

Let's create a `sayHello` step that takes the `name` parameter and echoes a simple message. This should be stored in the `vars/sayHello.groovy` file:

```
/**
 * Hello world step.
 */
def call(String name) {
    echo "Hello $name!"
}
```

> **Information**
>
> Human-readable descriptions for shared library steps can be stored in the `*.txt` files. In our example, we could add the `vars/sayHello.txt` file with the step documentation.

When the library code is done, we need to push it to the repository, for example, as a new GitHub project.

Configure the shared library in Jenkins

The next step is to register the shared library in Jenkins. We open **Manage Jenkins | Configure System** and find the **Global Pipeline Libraries** section. There, we can add the library giving it a chosen name, as follows:

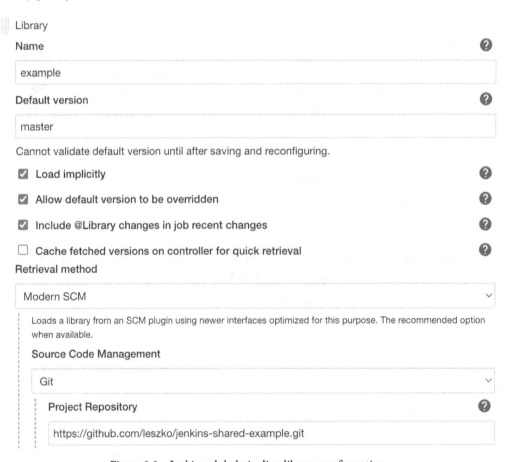

Figure 9.8 – Jenkins global pipeline library configuration

We specified the name under which the library is registered and the library repository address. Note that the latest version of the library will automatically be downloaded during the pipeline build.

> **Information**
>
> We showed importing the Groovy code as a *Global Shared Library*, but there are also other solutions. Read more at https://www.jenkins.io/doc/book/pipeline/shared-libraries/.

Using the shared library in a Jenkinsfile

Finally, we can use the shared library in a Jenkinsfile:

```
pipeline {
    agent any
    stages {
        stage("Hello stage") {
            steps {
                sayHello 'Rafal'
            }
        }
    }
}
```

> **Tip**
>
> If **Load implicitly** hadn't been checked in the Jenkins configuration, then we would need to add @Library('example') _ at the beginning of the Jenkinsfile script.

As you can see, we can use the Groovy code as a sayHello pipeline step. Obviously, after the pipeline build completes, we should see Hello Rafal! in the console output.

> **Information**
>
> Shared libraries are not limited to one step. Actually, with the power of the Groovy language, they can even act as templates for entire Jenkins pipelines.

After describing how to share the Jenkins pipeline code, let's also write a few words on rolling back deployments during the continuous delivery process.

Rolling back deployments

I remember the words of my colleague, a senior architect—*You don't need more QAs, you need a faster rollback*. While this statement is oversimplified and the QA team is often of great value, there is a lot of truth in this sentence. Think about it; if you introduce a bug in the production but roll it back soon after the first user reports an error, then usually, nothing bad happens. On the other hand, if production errors are rare but no rollback is applied, then the process to debug the production usually ends in long, sleepless nights and some dissatisfied users. That's why we need to think about the rollback strategy upfront while creating the Jenkins pipeline.

In the context of continuous delivery, there are two moments when the failure can happen:

- During the release process, in the pipeline execution
- After the pipeline build is completed, in production

The first scenario is pretty simple and harmless. It concerns a case when the application is already deployed to production but the next stage fails, for example, the smoke test. Then, all we need to do is execute a script in the `post` pipeline section for the `failure` case, which downgrades the production service to the older Docker image version. If we use blue-green deployment (as we will describe later in this chapter), the risk of any downtime is minimal, since we usually execute the load-balancer switch as the last pipeline stage after the smoke test.

The second scenario, in which we notice a production bug after the pipeline is successfully completed, is more difficult and requires a few words of comment. Here, the rule is that we should always release the rolled-back service using exactly the same process as the standard release. Otherwise, if we try to do something manually in a faster way, we are asking for trouble. Any non-repetitive task is risky, especially under stress when production is out of order.

> **Information**
>
> As a side note, if the pipeline completes successfully but there is a production bug, then it means that our tests are not good enough. So, the first thing after the rollback is to extend the unit/acceptance test suites with the corresponding scenarios.

The most common continuous delivery process is a single, fully automated pipeline that starts by checking out the code and ends with release to the production.

The following diagram shows how this works:

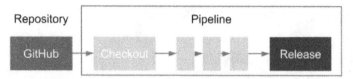

Figure 9.9 – Continuous delivery pipeline

We already presented the classic continuous delivery pipeline in this book. If the rollback should use exactly the same process, then all we need to do is revert the latest code change from the repository. As a result, the pipeline automatically builds, tests, and finally, releases the right version.

> **Information**
>
> Repository reverts and emergency fixes should never skip the testing stages in the pipeline, otherwise, we may end up with a release that is still not working correctly due to another issue that makes debugging even harder.

The solution is very simple and elegant. The only drawback is the downtime that we need to spend on the complete pipeline build. This downtime can be avoided if we use blue-green deployment or canary releases, in which cases, we only change the load balancer setting to address the healthy environment.

The rollback operation becomes far more complex in the case of orchestrated releases, during which many services are deployed at the same time. This is one of the reasons why orchestrated releases are treated as an anti-pattern, especially in the microservice world. The correct approach is to always maintain backward compatibility, at least for a time (as we showed for the database at the beginning of this chapter). Then, it's possible to release each service independently.

Adding manual steps

In general, the continuous delivery pipelines should be fully automated, triggered by a commit to the repository, and end after the release. Sometimes, however, we can't avoid having manual steps. The most common example is the release approval, which means that the process is fully automated, but there is a manual step to approve the new release. Another common example is manual tests. Some of them may exist because we operate on a legacy system; some others may occur when a test simply cannot be automated. No matter what the reason is, sometimes, there is no choice but to add a manual step.

Jenkins syntax offers an `input` keyword for manual steps:

```
stage("Release approval") {
    steps {
        input "Do you approve the release?"
    }
}
```

The pipeline will stop execution on the `input` step and wait until it's manually approved.

Remember that manual steps quickly become a bottleneck in the delivery process, and this is why they should always be treated as a solution that is inferior to complete automation.

> **Tip**
>
> It is sometimes useful to set a timeout for the input to avoid waiting forever for the manual interaction. After the configured time is elapsed, the whole pipeline is aborted.

We have covered a lot of important pipeline patterns; now, let's focus on different deployment release patterns.

Release patterns

In the last section, we discussed the Jenkins pipeline patterns used to speed up the build execution (parallel steps), help with the code reuse (shared libraries), limit the risk of production bugs (rollback), and deal with manual approvals (manual steps). This section will focus on the next group of patterns; this time, related to the release process. They are designed to reduce the risk of updating the production to a new software version.

We already described one of the release patterns, rolling updates, in *Chapter 6, Clustering with Kubernetes*. Here, we will present two more: blue-green deployment and canary releases.

> **Information**
>
> A very convenient way to use the release patterns in Kubernetes is to use the Istio service mesh. Read more at `https://istio.io/`.

Blue-green deployment

Blue-green deployment is a technique to reduce the downtime associated with the release. It concerns having two identical production environments—one called **green**, the other called **blue**—as presented in the following diagram:

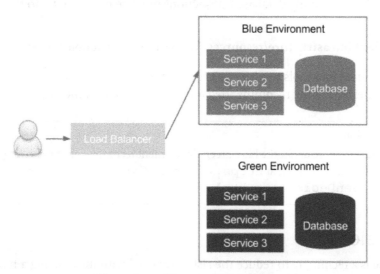

Figure 9.10 – Blue-green deployment

In the figure, the currently accessible environment is blue. If we want to make a new release, then we deploy everything to the green environment and, at the end of the release process, change the load balancer to the green environment. As a result, the user suddenly starts using the new version. The next time we want to make a release, we make changes to the blue environment and, in the end, we change the load balancer to blue. We proceed the same every time, switching from one environment to another.

> **Information**
>
> The blue-green deployment technique works correctly with two assumptions: environmental isolation and no orchestrated releases.

This solution provides the following benefits:

- **Zero downtime**: All the downtime, from the user perspective, is a moment of changing the load balance switch, which is negligible.

- **Rollback**: In order to roll back one version, it's enough to change back the load balance switch.

Note that the blue-green deployment must include the following:

- **Database**: Schema migrations can be tricky in case of a rollback, so it's worth using the patterns discussed at the beginning of this chapter.

- **Transactions**: Running database transactions must be handed over to the new database.

- **Redundant infrastructure/resources**: We need to have double the resources.

There are techniques and tools to overcome these challenges, so the blue-green deployment pattern is highly recommended and is widely used in the IT industry.

Information

You can read more about the blue-green deployment technique on the excellent blog from Martin Fowler, at `https://martinfowler.com/bliki/BlueGreenDeployment.html`.

Canary release

Canary release is a technique to reduce the risk associated with introducing a new version of the software. Similar to blue-green deployment, it uses two identical environments, as presented in the following diagram:

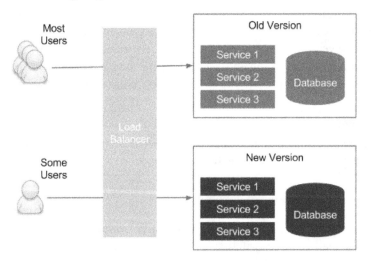

Figure 9.11 – Canary release

Also, similar to the blue-green deployment technique, the release process starts by deploying a new version in the environment that is currently unused. Here, however, the similarities end. The load balancer, instead of switching to the new environment, is set to link only a selected group of users to the new environment. The rest still use the old version. This way, a new version can be tested by some users and, in case of a bug, only a small group will be affected. After the testing period, all users are switched to the new version.

This approach has some great benefits:

- **Acceptance and performance testing**: If the acceptance and performance testing are difficult to run in the staging environment, then it's possible to test them in production, minimizing the impact on a small group of users.

- **Simple rollback**: If a new change causes a failure, then rolling back is done by switching all users back to the old version.

- **A/B testing**: If we are not sure whether the new version is better from the UX or the performance perspective, then it's possible to compare it with the old version.

Canary release shares the same drawbacks as the blue-green deployment. The additional challenge is that we have two production systems running at the same time. Nevertheless, canary release is an excellent technique used in most companies to help with the release and testing.

> **Information**
> You can read more about the canary release technique on Martin Fowler's blog, at `https://martinfowler.com/bliki/CanaryRelease.html`.

Working with legacy systems

Everything we have described so far applies to greenfield projects, for which setting up a continuous delivery pipeline is relatively simple.

Legacy systems are, however, far more challenging, because they usually depend on manual tests and manual deployment steps. In this section, we will walk through the recommended scenario to incrementally apply continuous delivery to a legacy system.

As the first step, I recommend reading a great book by Michael Feathers, *Working Effectively with Legacy Code*. His ideas on how to deal with testing, refactoring, and adding new features address most of the concerns about how to automate the delivery process for legacy systems.

> **Information**
>
> For many developers, it may be tempting to completely rewrite a legacy
> system rather than refactor it. While the idea is interesting from a developer's
> perspective, it is usually a bad business decision that results in a product failure.
> You can read more about the history of rewriting the Netscape browser in a
> brilliant blog post by Joel Spolsky, *Things You Should Never Do*, at `https://`
> `www.joelonsoftware.com/2000/04/06/things-you-`
> `should-never-do-part-i`.

The way to apply the continuous delivery process depends a lot on the current project's
automation, the technology used, the hardware infrastructure, and the current release
process. Usually, it can be split into three steps:

1. Automating build and deployment
2. Automating tests
3. Refactoring and introducing new features

Let's look at these in detail.

Automating build and deployment

The first step includes automating the deployment process. The good news is that
in most legacy systems that I have worked with, there was already some automation in
place (for example, in the form of shell scripts).

In any case, the activities for automated deployment include the following:

1. **Build and package**: Some automation usually already exists, in the form of
    Makefile, Ant, Maven, or any other build tool configuration, or a custom script.
2. **Database migration**: We need to start incrementally managing the database
    schema. This requires putting the current schema as an initial migration and
    making all the further changes with tools such as Flyway or Liquibase, as already
    described in this chapter.
3. **Deployment**: Even if the deployment process is fully manual, then there is usually a
    text/wiki page description that needs to be converted into an automated script.
4. **Repeatable configuration**: In legacy systems, configuration files are usually
    changed manually. We need to extract the configuration and use a configuration
    management tool, as described in *Chapter 7, Configuration Management with
    Ansible*.

After the preceding steps, we can put everything into a deployment pipeline and use it as an automated phase after a manual **user acceptance testing** (**UAT**) cycle.

From the process perspective, it's already worth starting to release more often. For example, if the release is yearly, try to do it quarterly, then monthly. The push for that factor will later result in faster-automated delivery adoption.

Automating tests

The next step, usually much more difficult, is to prepare the automated tests for the system. It requires communicating with the QA team in order to understand how they currently test the software so that we can move everything into an automated acceptance test suite. This phase requires two steps:

1. **Acceptance/sanity test suite**: We need to add automated tests that replace some of the regression activities of the QA team. Depending on the system, they can be provided as a black-box Selenium test or a Cucumber test.

2. **(Virtual) test environments**: At this point, we should already be thinking of the environments in which our tests would run. Usually, the best solution to save resources and limit the number of machines required is to virtualize the testing environment using Vagrant or Docker.

The ultimate goal is to have an automated acceptance test suite that will replace the whole UAT phase from the development cycle. Nevertheless, we can start with a sanity test that will check if the system is correct, from the regression perspective.

> **Information**
> While adding test scenarios, remember that the test suite should execute in a reasonable time. For sanity tests, it is usually less than 10 minutes.

Refactoring and introducing new features

When we have the fundamental regression testing suite (at a minimum), we are ready to add new features and refactor the old code. It's always better to do it in small pieces step by step, because refactoring everything at once usually ends up in chaos, and that leads to production failures (not related to any particular change).

This phase usually includes the following activities:

- **Refactoring**: The best place to start refactoring the old code is where the new features are expected. Starting this way prepares us for the new feature requests yet to come.

- **Rewrite**: If we plan to rewrite parts of the old code, we should start from the code that is the most difficult to test. This way, we can constantly increase the code coverage in our project.

- **Introducing new features**: During the new feature implementation, it's worth using the **feature toggle** pattern. Then, if anything bad happens, we can quickly turn off the new feature. The same pattern should also be used during refactoring.

> **Information**
>
> For this phase, it's worth reading a very good book by Martin Fowler, *Refactoring: Improving the Design of Existing Code*.

While touching on the old code, it's good to follow the rule to always add a passing unit test first, and only then change the code. With this approach, we can rely on automation to check that we don't accidentally change the business logic.

Understanding the human element

While introducing the automated delivery process to a legacy system, you may feel the human factor more than anywhere else. In order to automate the build process, we need to communicate well with the operations team, and they must be willing to share their knowledge. The same story applies to the manual QA team; they need to be involved in writing automated tests because only they know how to test the software. If you think about it, both the operations and QA teams need to contribute to the project that will later automate their work. At some point, they may realize that their future in the company is not stable and become less helpful. Many companies struggle with introducing the continuous delivery process because teams do not want to get involved enough.

In this section, we discussed how to approach legacy systems and the challenges they pose. If you are in the process of converting your project and organization to the continuous delivery approach, then you may want to take a look at the Continuous Delivery Maturity Model, which aims to give some structure to the process of adopting automated delivery.

Summary

This chapter has been a mixture of various continuous delivery aspects that were not previously covered. The key takeaways from the chapter are as follows:

- Databases are an essential part of most applications, and should, therefore, be included in the continuous delivery process.

- Database schema changes are stored in the version control system and managed by database migration tools.

- There are two types of database schema changes: backward-compatible and backward-incompatible. While the first type is simple, the second requires a bit of overhead (split to multiple migrations spread over time).

- A database should not be the central point of the whole system. The preferred solution is to provide each service with its own database.

- The delivery process should always be prepared for a rollback scenario.

- Three release patterns should always be considered: rolling updates, blue-green deployment, and canary release.

- Legacy systems can be converted to the continuous delivery process in small steps, rather than all at once.

Next, for the last part of the book, we will look into the best practices for your continuous delivery process.

Exercises

In this chapter, we covered various aspects of the continuous delivery process. Since practice makes perfect, we recommend the following exercises:

1. Use Flyway to create a non-backward-compatible change in the MySQL database:

    I.   Use the official Docker image, mysql, to start the database.

    II.   Configure Flyway with a proper database address, username, and password.

    III.   Create an initial migration that creates a USERS table with three columns: ID, EMAIL, and PASSWORD.

    IV.   Add sample data to the table.

    V.   Change the PASSWORD column to HASHED_PASSWORD, which will store the hashed passwords.

VI.   Split the non-backward-compatible change into three migrations, as described in this chapter.

VII. You can use MD5 or SHA for hashing.

VIII. Check that the database doesn't store any passwords in plain text as a result.

2. Create a Jenkins shared library with steps to build and unit test Gradle projects:

I.   Create a separate repository for the library.

II.   Create two files in the library: `gradleBuild.groovy` and `gradleTest.groovy`.

III. Write the appropriate `call` methods.

IV.   Add the library to Jenkins.

V.   Use the steps from the library in a pipeline.

Questions

To verify the knowledge from this chapter, please answer the following questions:

1. What are database (schema) migrations?
2. Can you name at least three database migration tools?
3. What are the main two types of changes to the database schema?
4. Why should one database not be shared between multiple services?
5. What is the difference between the test data for unit tests and integration/acceptance tests?
6. What Jenkins pipeline keyword do you use to make the steps run in parallel?
7. What are different methods to reuse Jenkins pipeline components?
8. What Jenkins pipeline keyword do you use to make a manual step?
9. What are the three release patterns mentioned in this chapter?

Further reading

To read more about the advanced aspects of the continuous delivery process, please refer to the following resources:

- *Databases as a Challenge for Continuous Delivery*: `https://phauer.com/2015/databases-challenge-continuous-delivery/`.

- *Zero Downtime Deployment with a Database*: `https://spring.io/blog/2016/05/31/zero-downtime-deployment-with-a-database`.

- *Canary Release*: `https://martinfowler.com/bliki/CanaryRelease.html`.

- *Blue-Green Deployment*: `https://martinfowler.com/bliki/BlueGreenDeployment.html`.

Best Practices

Thank you for reading this book. I hope you are ready to introduce the continuous delivery approach to your IT projects. By way of a final section for this book, I propose a list of the top 10 Continuous Delivery practices. Enjoy!

Practice 1 – Own the process within the team!

Own the entire process within the team, from receiving requirements to monitoring production. As was once remarked, "*A program running on the developer's machine makes no money.*" This is why it's important to have a small DevOps team that takes complete ownership of a product. Actually, that is the true meaning of **DevOps: Development and Operations**, from the beginning to the end:

- Own every stage of the Continuous Delivery pipeline: how to build the software, what the requirements are in acceptance tests, and how to release the product.

- Avoid having a pipeline expert! Every member of the team should be involved in creating the pipeline.

- Find a good way to share the current pipeline state (and the production monitoring) among team members. The most effective solution is big screens in the team space.

- If a developer, QA, and IT operations engineer are separate experts, then make sure they work together in one agile team. Separate teams based on expertise result in no one taking responsibility for the product.

- Remember that autonomy given to the team results in high job satisfaction and exceptional engagement. This leads to great products!

Practice 2 – Automate everything!

Automate everything, from business requirements (in the form of acceptance tests) to the deployment process. Manual descriptions, wiki pages with instruction steps, they all quickly become out of date and lead to tribal knowledge that makes the process slow, tedious, and unreliable. This, in turn, leads to a need for release rehearsals, and makes every deployment unique. Don't go down this path! As a rule, if you do anything for the second time, automate it:

- Eliminate all manual steps; they are a source of errors! The whole process must be repeatable and reliable.

- Don't ever make any changes directly in production! Use configuration management tools instead.

- Use precisely the same mechanism to deploy to every environment.

- Always include an automated smoke test to check whether the release was completed successfully.

- Use database schema migrations to automate database changes.

- Use automatic maintenance scripts for backup and cleanup. Don't forget to remove unused Docker images!

Practice 3 – Version everything!

Version everything: software source code, build scripts, automated tests, configuration management files, Continuous Delivery pipelines, monitoring scripts, binaries, and documentation; simply everything. Make your work task-based, where each task results in a commit to the repository, no matter whether it's related to requirement gathering, architecture design, configuration, or software development. A task starts on the agile board and ends up in the repository. This way, you maintain a single point of truth with the history and reasons for the changes:

- Be strict about version control. Version everything means everything!

- Keep the source code and configuration in the code repository, the binaries in the artifact repository, and the tasks in the agile issue tracking tool.

- Develop the Continuous Delivery pipeline as a code.

- Use database migrations and store them in a repository.

- Store documentation in the form of markdown files that can be version-controlled.

Practice 4 – Use business language for acceptance tests

Use business-facing language for acceptance tests to improve mutual communication and a common understanding of the requirements. Work closely with the product owner to create what Eric Evan called the *ubiquitous language*, a common dialect between the business and technology. Misunderstandings are the root cause of most project failures:

- Create a common language and use it inside the project.

- Use an acceptance testing framework, such as Cucumber or FitNesse, to help the business team understand and get them involved.

- Express business values inside acceptance tests, and don't forget about them during development. It's easy to spend too much time on unrelated topics!

- Improve and maintain acceptance tests so that they always act as regression tests.

- Make sure everyone is aware that a passing acceptance test suite means a green light from the business to release the software.

Practice 5 – Be ready to roll back

Be ready to roll back; sooner or later, you will need to do it. Remember, you don't need more QAs; you need a faster rollback. If anything goes wrong in production, the first thing you want to do is to play safe and come back to the last working version:

- Develop a rollback strategy and the process of what to do when the system is down.

- Split non-backward-compatible database changes into compatible ones.

- Always use the same process of delivery for rollbacks and standard releases.

- Consider introducing blue-green deployments or canary releases.

- Don't be afraid of bugs; the user won't leave you if you react quickly!

Practice 6 – Don't underestimate the impact of people

Don't underestimate the impact of people. They are usually way more important than tools. You won't automate delivery if the IT operations team won't help you. After all, they know the current process. The same applies to QAs, businesses, and everyone involved. Make them important and involved:

- Let QAs and IT operations be a part of the DevOps team. You need their knowledge and skills!

- Provide training to members who are currently doing manual activities so that they can move to automation.

- Favor informal communication and a flat structure of organization over hierarchy and orders. You won't do anything without goodwill!

Practice 7 – Incorporate traceability

Incorporate traceability for the delivery process and working system. There is nothing worse than a failure without any log messages. Monitor the number of requests, the latency, the load of production servers, the state of the Continuous Delivery pipeline, and everything you can think of that could help you to analyze your current software. Be proactive! At some point, you will need to check the stats and logs:

- Log pipeline activities! In the case of failure, notify the team with an informative message.

- Implement proper logging and monitoring of the running system.

- Use specialized tools for system monitoring, such as Kibana, Grafana, or Logmatic.io.

- Integrate production monitoring into your development ecosystem. Consider having big screens with the current production stats in the common team space.

Practice 8 – Integrate often

Integrate often; actually, all the time! As someone once said, "*Continuous is more often than you think.*" There is nothing more frustrating than resolving merge conflicts. Continuous integration is less about the tool and more about the team practice. Integrate the code into one code base at least a few times a day. Forget about long-lasting feature branches and a huge number of local changes. Trunk-based development and feature toggles for the win!

- Use trunk-based development and feature toggles instead of feature branches.

- If you need a branch or local changes, make sure that you integrate with the rest of the team at least once a day.

- Always keep the trunk healthy; make sure you run tests before you merge into the baseline.

- Run the pipeline after every commit to the repository for a faster feedback cycle.

Practice 9 – Only build binaries once

Build binaries only once, and run the same one on each of the environments, irrespective of whether they are in a form of Docker images or JAR packages; building only once eliminates the risk of differences introduced by various environments. It also saves time and resources:

- Build once, and pass the same binary between environments.

- Use an artifact repository to store and version binaries. Don't ever use the source code repository for that purpose.

- Externalize configurations and use a configuration management tool to introduce differences between environments.

Practice 10 – Release often

Release often, preferably following each commit to the repository. As the saying goes, *"If it hurts, do it more often."* Releasing as a daily routine makes the process predictable and calm. Stay away from being trapped in the rare release habit. That will only get worse and you will end up releasing once a year, having a three month preparation period!

- Rephrase your definition of done to *done means released*. Take ownership of the whole process!

- Use feature toggles to hide features that are still in progress from users.

- Use canary releases and quick rollback to reduce the risk of bugs in production.

- Adopt a zero-downtime deployment strategy to enable frequent releases.

With the final part of this book, we've covered the most important ideas and tooling around the Continuous Delivery process. I hope you found it valuable, and I wish you all the best in your Continuous Delivery journey!

Assessments

In the following pages, we will review all of the practice questions from each of the chapters in this book and provide the correct answers.

Chapter 1: Introducing Continuous Delivery

1. Development, quality assurance, operations.
2. Continuous integration, automated acceptance testing, configuration management.
3. Fast delivery, fast feedback cycle, low-risk releases, flexible release options.
4. Unit tests, integration tests, acceptance tests, non-functional tests (performance, security, scalability, and so on).
5. Unit tests, because they are cheap to create/maintain and quick to execute.
6. DevOps is the idea of combining the areas of development, quality assurance, and operations into one team (or person). Thanks to automation, it's possible to provide the product from A to Z.
7. Docker, Jenkins, Ansible, Terraform, Git, Java, Spring Boot, Gradle, Cucumber, Kubernetes.

Chapter 2: Introducing Docker

1. Containerization does not emulate the whole operating system; it uses the host operating system instead.
2. The benefits of providing an application as a Docker image are as follows:

    I. **No issues with dependencies**: The application is provided together with its dependencies.

    II. **Isolation**: The application is isolated from the other applications running on the same machine.

    III. **Portability**: The application runs everywhere, no matter which environment dependencies are present.

3. No, the Docker daemon can run natively only on Linux machines. However, there are well-integrated virtual environments for both Windows and Mac.

4. A Docker image is a stateless, serialized collection of files and the recipe of how to use them; a Docker container is a running instance of the Docker image.

5. A Docker image is built on top of another Docker image, which makes the layered structure. This mechanism is user-friendly and saves bandwidth and storage.

6. Docker Commit and Dockerfile.

7. `docker build`.

8. `docker run`.

9. Publishing a port means that the host's port is forwarded to the container's port.

10. A Docker volume is the Docker host's directory mounted inside the container.

Chapter 3: Configuring Jenkins

1. Yes, and the image name is `jenkins/jenkins`.

2. A Jenkins master is the main instance that schedules tasks and provides the web interface, while a Jenkins agent (slave) is the additional instance that's only dedicated to executing jobs.

3. Vertical scaling means adding more resources to the machine while the load increases. Horizontal scaling means adding more machines while the load increases.

4. SSH and Java Web Start.

5. A permanent agent is the simplest solution, and it means creating a static server with all the environments prepared to execute a Jenkins job. On the other hand, a permanent Docker agent is more flexible; it provides the Docker daemon, and all the jobs are executed inside Docker containers.

6. In case you use dynamically provisioned Docker agents and the standard ones (available on the internet) do not provide the execution environment you need.

7. When your organization needs some templated Jenkins to be used by different teams.

8. Blue Ocean is a Jenkins plugin that provides a more modern Jenkins web interface.

Chapter 4: Continuous Integration Pipeline

1. A pipeline is a sequence of automated operations that usually represents a part of the software delivery and quality assurance process.

2. A step is a single automated operation, while a stage is a logical grouping of steps used to visualize the Jenkins pipeline process.

3. The `post` section defines a series of one or more step instructions that are run at the end of the pipeline build.

4. Checkout, compile, and unit test.

5. Jenkinsfile is a file with the Jenkins pipeline definition (usually stored together with the source code in the repository).

6. The code coverage stage is responsible for checking whether the source code is well covered with (unit) tests.

7. An external trigger is a call from an external repository (such as GitHub) to the Jenkins master, while Polling SCM is a periodic call from the Jenkins master to the external repository.

8. Email, group chat, build radiators, sms, rss feed.

9. Trunk-based workflow, branching workflow, and forking workflow.

10. A feature toggle is a technique that is used to disable the feature for users but enable it for developers while testing. Feature toggles are essentially variables used in conditional statements.

Chapter 5: Automated Acceptance Testing

1. Docker Registry is a stateless application server that stores Docker images.

2. Docker Hub is the best-known public Docker registry.

3. The convention is `<registry_address>/<image_name>:<tag>`.

4. The staging environment is the preproduction environment dedicated to integration and acceptance testing.

5. The following commands: `docker build`, `docker login`, and `docker push`.

6. They allow us to specify tests in a human-readable format, which helps with collaboration between businesses and developers.

7. Acceptance criteria (feature scenario specification), step definitions, test runner.

8. Acceptance test-driven development is a development methodology (seen as an extension of TDD) that says to always start the development process from the (failing) acceptance tests.

Chapter 6: Clustering with Kubernetes

1. A server cluster is a set of connected computers that work together in such a way that they can be used similarly within a single system.

2. Kubernetes Node is just a worker, that is, a host that runs containers. Kubernetes Control Plane Master is responsible for everything else (providing the Kubernetes API, Pod orchestration, and more).

3. Microsoft Azure, Google Cloud Platform, and Amazon Web Services.

4. Deployment is a Kubernetes resource that's responsible for Pod orchestration (creating, terminating, and more). Service is an (internal) load balancer that provides a way to expose Pods.

5. `kubectl scale`.

6. Docker Swarm and Mesos.

Chapter 7: Configuration Management with Ansible

1. Configuration management is the process of controlling the configuration changes in a way such that the system maintains integrity over time.

2. Agentless means that you don't need to install any special tool (an agent or daemon) in the server that is being managed.

3. Ansible, Chef, and Puppet.

4. An inventory is a file that contains a list of servers that are managed by Ansible.

5. An ad hoc command is a single command that is executed on servers, and playbooks are the entire configurations (sets of scripts) that are executed on servers.

6. An Ansible role is a well-structured playbook prepared to be included in the playbooks.

7. Ansible Galaxy is a store (repository) for Ansible roles.

8. Infrastructure as code is the process of managing and provisioning computing resources instead of physical hardware configurations.

9. Terraform, AWS CloudFormation, Azure Resource Manager, Google Cloud Deployment Manager, Ansible, Chef, Puppet, Pulumi, Vagrant.

Chapter 8: Continuous Delivery Pipeline

1. Production, staging, QA, development.

2. Staging is the preproduction environment used to test software before the release; QA is a separate environment used by the QA team and the dependent applications.

3. Performance, load, stress, scalability, endurance, security, maintainability, recovery.

4. No, but it should be explicit which are part of the pipeline, and which are not (and for those that are not, there should still be some automation and monitoring around).

5. Semantic versioning, timestamp-based, hash-based.

6. A smoke test is a very small subset of acceptance tests whose only purpose is to check that the release process is completed successfully.

Chapter 9: Advanced Continuous Delivery

1. Database schema migration is a process of incremental changes to the relational database structure.

2. Flyway, Liquibase, Rail migrations (from Ruby on Rails), Redgate, Optima database administrator.

3. Backward-compatible and non-backward-compatible.

4. If one database is shared between multiple services, then each database change must be compatible with all services, which makes changes very difficult to initiate.

5. Unit tests do not require the preparation of any special data; data is in memory and prepared by developers; integration/acceptance tests require the preparation of special data that is similar to production data.

6. Parallel.

7. Build parameters and shared libraries.

8. Input.

9. Rolling updates, blue-green deployment, and canary release.

Index

`Packt.com`

Subscribe to our online digital library for full access to over 7,000 books and videos, as well as industry leading tools to help you plan your personal development and advance your career. For more information, please visit our website.

Why subscribe?

- Spend less time learning and more time coding with practical eBooks and Videos from over 4,000 industry professionals

- Improve your learning with Skill Plans built especially for you

- Get a free eBook or video every month

- Fully searchable for easy access to vital information

- Copy and paste, print, and bookmark content

Did you know that Packt offers eBook versions of every book published, with PDF and ePub files available? You can upgrade to the eBook version at `packt.com` and as a print book customer, you are entitled to a discount on the eBook copy. Get in touch with us at `customercare@packtpub.com` for more details.

At `www.packt.com`, you can also read a collection of free technical articles, sign up for a range of free newsletters, and receive exclusive discounts and offers on Packt books and eBooks.

Other Books You May Enjoy

If you enjoyed this book, you may be interested in these other books by Packt:

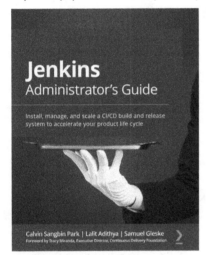

Jenkins Administrator's Guide

Calvin Sangbin Park, Lalit Adithya, Samuel Gleske

ISBN: 9781838824327

- Set up a production-grade Jenkins instance on AWS and on-premises
- Create continuous integration and continuous delivery (CI/CD) pipelines triggered by GitHub pull request events
- Use Jenkins Configuration as Code to codify a Jenkins setup
- Backup and restore configurations and plan for disaster recovery
- Plan, communicate, execute, and roll back upgrade scenarios
- Identify and remove common bottlenecks in scaling Jenkins
- Use Shared Libraries to develop helper functions and create new DSLs

Mastering Docker, Fourth Edition

Russ McKendrick

ISBN: 9781839216572

- Get to grips with essential Docker components and concepts
- Discover the best ways to build, store, and distribute container images
- Understand how Docker can fit into your development workflow
- Secure your containers and files with Docker's security features
- Explore first-party and third-party cluster tools and plugins
- Launch and manage your Kubernetes clusters in major public clouds

Packt is searching for authors like you

If you're interested in becoming an author for Packt, please visit `authors.packtpub.com` and apply today. We have worked with thousands of developers and tech professionals, just like you, to help them share their insight with the global tech community. You can make a general application, apply for a specific hot topic that we are recruiting an author for, or submit your own idea.

Share Your Thoughts

Now you've finished *Continuous Delivery with Docker and Jenkins*, we'd love to hear your thoughts! Scan the QR code below to go straight to the Amazon review page for this book and share your feedback or leave a review on the site that you purchased it from.

`https://packt.link/r/1803237481`

Your review is important to us and the tech community and will help us make sure we're delivering excellent quality content.

www.ingramcontent.com/pod-product-compliance
Lightning Source LLC
Chambersburg PA
CBHW062049050326
40690CB00016B/3030